Establishing a Safety-First Corporate Culture in Your Organization

An Integrated Approach for Safety Professionals and Safety Committees

D1465859

Establishing a Safety-First Corporate Culture in Your Organization

An Integrated Approach for Safety Professionals and Safety Committees

David L. Goetsch

Prentice Hall

Boston   Columbus   Indianapolis   New York   San Francisco   Upper Saddle River   Amsterdam
Cape Town   Dubai   London   Madrid   Milan   Munich   Paris   Montreal   Toronto   Delhi
Mexico City   Sao Paulo   Sydney   Hong Kong   Seoul   Singapore   Taipei   Tokyo

Vice President and Executive Publisher: Vernon R. Anthony
Acquisitions Editor: David Ploskonka
Editorial Assistant: Nancy Kesterson
Production Manager: Wanda Rockwell
Creative Director: Jayne Conte
Cover Designer and Art: Aaron Dixon
Director of Marketing: David Gesell
Senior Marketing Coordinator: Alicia Wozniak

This book was set in 10/12 Times by Aptrara®, Inc., and was printed and bound by Courier. The cover was printed by Courier.

Library of Congress Cataloging-in-Publication Data

Goetsch, David L.
  Establishing a safety-first corporate culture in your organization: an integrated approach for safety professionals and safety committees / David L. Goetsch.
      p. cm.
  ISBN-13: 978-0-13-502597-0
  ISBN-10: 0-13-502597-4
  1. Industrial safety—Management. 2. Corporate culture. I. Title.
  T55.G5849 2010
  658.3'82—dc22

                                                                                    2009017697

10 9 8 7 6 5 4 3 2 1

Prentice Hall
is an imprint of

www.pearsonhighered.com

ISBN-10:   0-13-502597-4
ISBN-13: 978-0-13-502597-0

Acknowledgments

The author would like to thank the following reviewers for their time and assistance with the text: Robert L. Hobbs, *Gulf Power Company & Southern Company's Safety and Health Council;* Larry Leiman, *Safety Consulting & Training, Inc.;* Donna G. Miller, *Okaloosa County Board of County Commissioners;* John Renzelman, *Wayne State College;* and Joyce Szilvasy, *Choctawatchee Electric Cooperative, Inc.*

Contents

About the Author

David L. Goetsch is Vice President of Northwest Florida State College and Provost of the College's joint campus with the University of West Florida in Fort Walton Beach, Florida. He is also a professor of safety, quality, and environmental management.

Dr. Goetsch administers the state of Florida's Center for Manufacturing Competitiveness that is located on this campus and the Leadership Institute, which is the corporate training Center of Northwest Florida State College. In addition, Dr. Goetsch is president and CEO of the Center for Organizational Excellence (COE), a private consulting firm dedicated to helping organizations achieve peak performance and continual improvement. Dr. Goetsch is a nationally recognized speaker and corporate trainer on the topics of safety, quality, and competitiveness.

Introduction

Establishing a Safety-First Corporate Culture

When a workplace safety tragedy occurs, invariably the term *safety culture* will surface in the investigations, discussions, debates, and reports that follow. More to the point, the organization in question is typically said to lack a safety culture or a culture of safety. Taking no exception with the point being made by those who use this term, *safety culture* is a bit of a misnomer. To use this term is to imply that safety is a stand-alone, nonintegrated appendage that can have its own culture separate from the broader corporate culture of the organization in question. This, of course, is not possible.

The commitment, attitudes, and practices of an organization as they relate to safety are all part of its broader corporate culture. For this reason, I prefer the term *safety-first corporate culture.* Using this term makes the important point that safety is, as it must be, a fully integrated element of an organization's broader corporate culture—an element that must come first on the organization's list of priorities. Establishing a safety-first corporate culture in your organization is the right thing to do for a number of reasons, all of which are explained in this book.

What Is a Safety-First Corporate Culture?

The term *corporate culture* refers to the tacit assumptions, beliefs, values, attitudes, expectations, and behaviors that are widely shared and accepted in an organization. An organization's corporate culture determines how employees approach their work, interact with each other, treat customers and suppliers, behave when the boss is not looking, and do their jobs.

Many organizations, particularly those that must compete in the global marketplace, have come to view safety as a corporate culture issue. For example, Voisey's Bay Nickel Company Limited says this about workplace safety:

> A safe workplace cannot be created and maintained through regulations alone. Safety targets cannot be met without a *strong culture of safety* [author's emphasis], reflected in the day-to-day behavior of all workers.[1]

Ironore Company of Canada provides another example of a global organization that views workplace safety as a corporate culture issue, as can be seen in the following statement:

> The people that go to work everyday at IOC expect to do so in a safe and healthy work environment. This is not an option but a minimum requirement. *Creating a culture* [author's emphasis] in which everyone contributes proactively to their own safety and the safety of their co-workers requires many things.[2]

With this background, a safety-first corporate culture can be defined as follows:

> *A corporate culture in which the tacit assumptions, beliefs, values, attitudes, expectations, and behaviors that are widely shared and accepted in an organization support the establishment and maintenance of a safe and healthy work environment for all personnel and other stakeholders.*

This definition makes the critical point that workplace safety is an integral part of an organization's overall corporate culture, not a separate and distinct component that stands alone. Workplace safety should be so integrated into an organization's corporate culture that it is viewed by key decision makers as part of the organization's competitive strategy for winning in the global marketplace.

For example, the full integration of workplace safety is the point being made in a report prepared by Stillwater Mining Company that says: "Without such an integrated approach, changes might be made for the sake of production or quality which could have a negative impact on the safety and health aspects."[3] For workplace safety to be fully integrated into an organization's processes and daily operations, it must be a fully integrated element of the organization's corporate culture.

What Does a Safety-First Corporate Culture Look Like?

If the prevailing tacit assumptions, beliefs, values, attitudes, expectations, and behaviors in an organization support safety, certain evidence should be apparent to an informed observer. Said another way, there are predictable, observable actions that organizations with a safety-first corporate culture take to establish and maintain such a culture. What follows are several things organizations that have established and maintain a safety-first corporate culture actually do that sets them apart from other organizations:

- Ensure that all personnel at all levels are committed to workplace safety and health.
- Ensure that personnel at all levels expect appropriate safety-first attitudes and practices from each other.
- Ensure that all personnel in positions of authority from executives through team leaders are consistently positive role models of appropriate safety-first attitudes and practices.
- Provide all new personnel with a comprehensive orientation of which safety-first attitudes and practices are an important component.
- Provide mentors to new personnel, who are responsible for, among other things, teaching that the safe way is the right way.
- Provide comprehensive, ongoing safety training for all personnel at all levels.
- Make safety part of the organization's team-building process.

- Monitor and evaluate the safety-first attitudes and practices of personnel at all levels.
- Periodically assess the safety aspects of the corporate culture and take corrective action wherever necessary.

If having a safety-first corporate culture is a test for organizations—and it is in today's hyper-competitive global marketplace—these criteria are the minimum needed for a passing score. Doing the things listed above amounts to building a supportive infrastructure that will undergird your organization's efforts to establish and maintain a safety-first corporate culture. Broadly speaking, these actions—when taken together—form the foundation upon which an organization builds a safety-first corporate culture.

Importance of a Safety-First Corporate Culture

When asked to summarize as briefly as possible why it is so important for organizations to establish and maintain a safety-first corporate culture, four words come immediately to mind. These words are "ethics," "compliance," "competition," and "cost." The first three rationale—ethics, compliance, and competition—all have imbedded within them the fourth element, cost. Failings in any of the first three areas can increase the cost of doing business for an organization.

Ethical responsibility has always been an important part of the rationale for workplace safety and health. Regulatory compliance became an important part of the rationale with passage of the Occupational Safety and Health Act (OSH Act) in 1970 and subsequent federal legislation such as the Federal Mine Safety Act, Clean Air Act, and Superfund Amendments. Competitiveness has always been an important rationale for maintaining a safe and healthy workplace, but, in recent years, the demands of global competition have increased the importance of this rationale for workplace safety.

From a business perspective, global competition has emerged as at least a coequal factor in the rationale for workplace safety because the ability to win consistently in the global marketplace has become a prerequisite for survival. An organization's image, employee morale, and product quality can be damaged by unethical safety lapses. Its bottom line can be undermined by fines levied for regulatory violations. But an organization's very survival is threatened if it cannot compete in the global arena. An organization that fails to provide a work environment that is conducive to peak performance, continual improvement, and superior value will not be able to compete on a global scale for very long.

The competitive value of a safe and healthy workplace is an important concept for safety professionals to understand and, in turn, help key decision makers in their organizations to understand. After all, it is often the bottom-line pressures of competition that cause organizations to cut corners when it comes to workplace safety and health. Such an approach is always shortsighted, often costly, and potentially damaging from the perspective of long-term competitiveness.

Competition has always been an issue for organizations. For example, before Toyota, Honda, and a host of other foreign automobile manufacturers became major players in the global marketplace, Ford and General Motors already had each other as competitors. But the concept of globalization has increased the intensity of competition, which has, in turn, increased the need for a safe and healthy work environment. To thrive or just survive in a globally competitive environment, organizations must be able to provide their customers with superior value, a concept achieved by providing superior quality, superior cost, and superior service.

To provide superior value, organizations must be innovative, flexible, lean, productive, and committed to continually improving their people, products, processes, and performance. This challenge is difficult enough under even the best circumstances. It simply cannot be done in an unsafe or unhealthy work environment where employee morale and organizational resources are continually drained off by accidents and injuries.

How to Establish a Safety-First Corporate Culture

What follows is a 10-step model for establishing a safety-first corporate culture in your organization. The model cannot be successfully implemented without the commitment of executives, managers, and supervisors. However, the key players in implementing this model are the organization's safety professionals and the safety committee. Safety professionals and the safety committee facilitate, moderate, and advocate on behalf of the model and its effective implementation. This means they have to lead, pull, push, prod, inform, convince, build consensus, defend, and—most important—persevere. Establishing a safety-first corporate culture will be a challenge in some organizations, but the payoff is worth it for the organizations and safety professionals, who invest the time, energy, and expertise necessary to make it happen. The steps for establishing a safety-first corporate culture are as follows:

- *Gain commitment* to a safety-first corporate culture.
- *Expect* safety-first attitudes and practices.
- *Role model* the expected safety-first attitudes and practices.
- *Orient* personnel to the expected safety-first attitudes and practices.
- *Mentor* personnel on the expected safety-first attitudes and practices.
- *Train* personnel on safety-first attitudes and practices.
- *Build teams* around the concept of safety.
- *Monitor/evaluate* safety-first attitudes and practices constantly.
- *Reinforce/integrate* safety-first attitudes and practices.
- *Assess* the safety aspects of the corporate culture periodically.

Each of these steps is the subject of a subsequent chapter in this book. In the chapters that follow, each respective step in the model is explained in detail.

Endnotes

1. Voisey's Bay Nickel Company Limited. "Social Responsibility Report," retrieved June 2008 from http://www.vbnc.com/Social Responsibility2005/safety/index.htm

2. Ironore Company of Canada, Retrieved June 2008 from http://www.ironore.ca/main/index.php?sec=4&loc=44&lng=EN

3. Stillwater Mining Company. "Safety Initiatives and Strategy," retrieved June 2008 from http://www.stillwatermining.com/CorporateResponsibility/Safety/Initiatives.html

Chapter 1

Gain Commitment to a Safety-First Corporate Culture

Major Topics

- Competitiveness rationale
- Cost rationale
- Ethics rationale
- Compliance rationale
- Persuasion strategies for safety professionals

The first step in establishing a safety-first corporate culture is to gain the commitment of key personnel in an organization. Eventually, the effort will require the commitment of all personnel, but initially, it is important to focus on gaining the commitment of key executives, managers, and supervisors. To gain their commitment, it will be necessary to ensure that these key personnel understand and accept the need for a safety-first corporate culture.

This means that safety professionals must fully understand the need for a safety-first corporate culture themselves and be able to articulate the need in a clear, compelling, and succinct manner. Safety professionals within an organization must use their understanding of the value of a safety-first corporate culture to develop a compelling rationale for the concept and employ this rationale to gain the commitment of other key personnel. Those other key personnel include the organization's corporate executive officer (CEO), executives, managers, supervisors, and all members of the safety committee.

Gaining the commitment of key decision makers in an organization can be the most difficult challenge safety professionals will face in implementing the model explained in this book. However, it is a challenge that must be met if the model is going to work. This is the bad news. The good news is that by using the reasons provided in this chapter, coupled with effective persuasion strategies and persistent effort, this challenge can be met.

In my experience, decision makers in an organization come to understand the need for a safety-first corporate culture in one of the following two ways: (1) as the result of a major incident or disaster that forces them to focus on safety or (2) due to the persistent effort and effective persuasion of dedicated safety professionals, who take advantage of every opportunity to show key personnel in their organizations the connection between safety and success in today's hypercompetitive, heavily regulated, and potentially litigious business environment.

In this regard, there are four major reasons why establishing and maintaining a safety-first corporate culture is essential in today's business environment: (1) the need to be competitive, (2) the need to hold down the cost of doing business (which is really embedded in the other three reasons), (3) the demands of corporate ethics, and (4) the need for regulatory compliance. The rationale for a safety-first corporate culture as it relates to each of these factors is explained in the sections of this chapter that follow.

Competitiveness Rationale

To survive and thrive in today's global marketplace, organizations must be able to compete at a world-class level because it takes world-class performance to win. Ulrich Becker, director of occupational safety and health for Germany's Federal Ministry of Economics and Labor, has this to say about safety and global competitiveness:

> In order to be able to compete successfully in today's market, more than ever companies need to recognize the expertise, experience and knowledge of employees, and to make systematic and efficient use of it and to promote it. The investment made today in human resources is decisive for the performance, the innovative capacities, and thus the future of a firm. Employee commitment, motivation, health and satisfaction are key factors underlying the competitiveness and profitability of a company and the sustainable success of an economy.[1]

Competing in the global business arena is like running an Olympic race that never ends. You might be ahead of the competition at the moment, but the race is not over. In fact, it never is. This is why competing in the global marketplace has been referred to as running a race that has no finish line.

Competitiveness, from the perspective of business and industry, is the ability to consistently succeed in the marketplace regardless of whether the marketplace is local, regional, national, or global. There is a close correlation between safety and competitiveness. A safe working environment allows employees to focus on achieving peak performance and continual improvement—both of which enhance competitiveness. A study conducted by the European Agency for Safety and Health at Work found that "there is a strong relationship: the higher the OSH (occupational safety and health) standards, the higher the productivity. In some cases, a good safety record can even be used to predict future profitability."[2]

Three concepts are fundamental to a theoretical understanding of competitiveness: (1) superior value, (2) peak performance, and (3) continual improvement. To win consistently in a competitive marketplace, organizations must provide their customers with superior value, a combination of superior quality, superior cost, and superior service. To consistently provide superior value, organizations need their personnel and processes to constantly operate at peak levels (peak performance) and get better all the time (continual improvement).

To achieve peak performance and ensure continual improvement, organizations must do at least all of the following:

- Recruit and retain high-value employees
- Provide ongoing training for all personnel
- Acquire productivity-enhancing technologies and continually update those technologies
- Undertake ongoing process improvement initiatives
- Undertake ongoing customer/market research
- Undertake ongoing product improvement initiatives
- Establish mutually supportive supplier partnerships

However, a safe work environment is a prerequisite for peak performance and continual improvement. Not one of the competitiveness imperatives mentioned above can be consistently achieved in a hazardous environment in which (1) employees feel threatened by unsafe conditions; (2) morale is sapped by accidents, incidents, and near misses; and (3) financial resources are steadily drained off to pay the counterproductive costs of accidents and injuries. In short, employees cannot focus on peak performance and continual improvement when they are distracted by hazardous working conditions.

Every tactic for enhancing competitiveness contained in the previous list is dependent on organizations' most finite resource, money. An unalterable truth is that organizations can use their limited financial resources to enhance competitiveness, or they can use them to pay the counterproductive costs of accidents, injuries, and noncompliance. Said another way, organizations that maintain a safe and healthy work environment are better able to invest in continually improving their people, processes, and products. Those that turn a blind eye to hazards in the work environment typically see the funds they need to invest in becoming more competitive siphoned off by the costs associated with accidents, injuries, and noncompliance. Decision makers who are responsible for maximizing their organization's competitiveness need to understand that now, more than at any time in the past, the most productive work environment is a safe work environment.

Cost Rationale

The costs associated with an unsafe workplace can undermine an organization's ability to compete, as explained in the previous section. These costs are principally financial, but they also include indirect costs such as those associated with poor employee morale and a damaged corporate image, both of which eventually translate into financial costs, as do all indirect costs.

Financial Costs of Accidents and Injuries

The direct costs of workplace accidents in the United States are well known to safety professionals. They approach $50 billion annually, and this figure does not include the fines assessed against organizations as the result of regulatory noncompliance. Here are just a few examples for the sake of illustration. As the result of a chemical explosion at its petrochemical plant in Channelview, Texas, Arco Chemical Company was once fined

$3.48 million. BASF Corporation once agreed to pay $1.06 million to settle Occupational Safety and Health Administration (OSHA) citations associated with an explosion at a Cincinnati chemical plant that caused the death of 2 personnel and injuries to 17 others. These were just two examples of the types of cases safety professionals read about regularly in their professional literature and might use to help the safety committee gain the commitment of decision makers in their organization for a safety-first corporate culture. The literature for safety professionals is replete with similar cases.

The figure of $50 billion, stated previously, includes direct costs accruing from the following. Keep in mind that there are also indirect costs associated with each of these areas:

- Production stoppages and delays
- Wages paid while the employee is off from work
- Medical expenses
- Insurance costs (including workers' compensation)
- Property damage
- Fire losses
- Survivor benefits
- Death and burial benefits
- Litigation
- Facility remediation
- Fines assessed for noncompliance
- Turnover
- Wages lost

As a safety professional, you can help decision makers understand the importance of a safety-first corporate culture by identifying these costs on an annual basis specifically for your organization. In my experience, executives and managers are typically shocked to learn how much their organization spends on just the direct costs in each of these areas.

Employee Morale Costs of Accidents and Injuries

Although it is difficult to quantify, many safety professionals believe that the damage done to employee morale is one of the highest costs of accidents and injuries. Employee morale is a less tangible factor than medical and insurance costs. However, it is widely accepted among management professionals that few factors affect productivity more profoundly than employee morale. Employees with low morale do not perform at peak levels, nor do they strive to improve continually. In other words, they do not commit to their organization's success.

In fact, when morale plummets, employees often show their dissatisfaction with their feet—they leave. Turnover is one of the most costly by-products of low employee morale. In a study by the European Agency for Safety and Health at Work, one company reported cutting its turnover rate down to half of the national average by implementing workplace safety measures.[3] This company then enjoyed the inherent cost savings associated with decreased turnover.

The cost of turnover explains why so much money is spent in corporate America every year to help managers and supervisors learn different strategies for improving employee morale. Few things are as detrimental to employee morale as seeing a fellow employee injured or killed on the job. Whenever this happens, other employees silently think, "That could have been me." Then, they begin to think about the family members of the injured or killed employee and wonder what will happen to them. This, in turn, leads to resentment. Employees whose concentration is distracted by fear for their personal safety and whose motivation is undermined by resentment over accidents and injuries will not perform at peak levels, will not do what is necessary to continually improve, and will not go the extra mile to help their employer become more competitive.

This contention was borne out in a survey in which employees were asked to identify factors that affected their morale, motivation, and performance on the job.[4] More than 90 percent of the respondents identified "quality of the work environment" as an important factor in determining their morale, motivation, and performance. An even more interesting study would be one to determine the cost to unsafe organizations of the resentment that builds up in the hearts and minds of employees who do not leave because of accidents and injuries. Rather, they respond to them by doing as little as possible in an attempt to gain a measure of retribution against their employer. I am not sure how such a study would be conducted or whether it even could be. However, my sense of it is that such a study would reveal that the cost to employers of this simmering resentment is equal to or even greater than the cost of turnover.

Corporate Image-Related Costs of Accidents and Injuries

In a competitive environment, an organization must be concerned about its corporate image. A positive corporate image can help make an organization more competitive, while a negative corporate image can have the opposite effect. For example, an image of being concerned about the safety and health of personnel can help an organization attract and retain the best employees. A reputation for being concerned about product safety can help the organization attract more customers. An image of being concerned about the environment can make the organization a more welcome corporate neighbor.

In contrast, an organization that develops a negative corporate image that conveys a lack of concern for the safety of its personnel, customers, and the environment will find it increasingly difficult to stay afloat in today's hypercompetitive business environment. This is more so because major accidents or incidents are likely to receive a great deal of attention as media outlets worldwide compete for a finite number of newsworthy stories. Further, a positive corporate image is a fragile phenomenon: Once earned, it is easily lost. It can never be taken for granted, but must be reearned every day. Just one major safety or health disaster, environmental catastrophe, or product safety tragedy—if poorly handled— can permanently undermine an organization's corporate image.

Ethics Rationale

When the subject of business ethics comes up in a conversation, the tendency is to think of financial fraud and other money-related improprieties. Financial integrity is, of course, an important ethical issue for businesses. However, the quality of the work environment is no less important an ethical issue. Organizations have an ethical obligation to provide a safe and healthy work environment for employees, safe products for customers, and safe disposal of their hazardous by-products for the communities in which they are located. Failure to do any one of these is a serious violation of an organization's corporate social responsibility.

The profusion of major ethical lapses associated with well-known corporations over the past two decades has now led many organizations to adopt a statement of corporate social responsibility and provide ethics training that is mandatory for all personnel. The challenge for safety professionals is to make sure that workplace, product, and environmental safety are all included in any code of corporate social responsibility adopted by their organizations.

It is important to ensure that workplace safety in particular is included in (1) the core values of an organization's strategic plan, (2) the organization's statement of corporate social responsibility, (3) all job descriptions, (4) all performance appraisal instruments, (5) team charters, and (6) corporate training programs. It is also important to have a corporate safety policy. All of these are observable evidence of a commitment to safety. Of course, words on paper may be nothing more than words. However, if the organization records its safety-related expectations in these various ways and matches words with action, a safety-first corporate culture can be established and maintained.

SAFETY-FIRST FACT

Safety Must Be More Than Just an Obligation

"Facilitating a safety culture is the most important task we have. There is no CHELCO without its employees; each of those employees has the right to work in a safe environment. We strive to go beyond the regulatory requirements and provide the tools, training, and incentives to ensure that our employees view safety as more than an obligation, but adopt it as a way of life."

J. E. "Gene" Smith, CEO
Choctawhatchee Electric Cooperative (CHELCO)

Source: E-mail to the author dated June 11, 2008

Compliance Rationale

The compliance rationale is one that safety professionals understand well because they live with it daily. Safety professionals represent organizations' front line when it comes to ensuring that organizations comply with all applicable local, state, and federal standards and guidelines relating to workplace safety and health. In terms of compliance, the principal driver is the OSHA. Another important player is the National Institute for Occupational Safety and Health (NIOSH). Both of these federal agencies were established by the Occupational Safety and Health Act of 1970 (OSH Act).

The OSH Act applies to most organizations—not all, but most. Organizations that fall under OSHA's umbrella are responsible for knowing which standards apply to them, effectively implementing those standards, properly posting required information, and appropriately reporting on workplace safety and health issues as specified by OSHA. Responsibility for making sure that organizations comply with all applicable OSHA and NIOSH standards, as well as any others that apply, is typically delegated to organizations' safety professionals. It is, of course, fitting and appropriate to delegate the authority for leading and managing organizations' safety programs to safety professionals. After all, they are the professionals and it's their job. However, key decision makers need to understand that delegating this authority does not absolve them of responsibility.

A challenge for safety professionals is to ensure that all key decision makers in their organizations understand their roles relating to safety and how costly failure to comply with applicable standards can be. For example, OSHA is empowered to levy citations and accompanying monetary penalties (fines) in an escalating list of categories, including (1) other-than-serious violations, (2) serious violations, (3) willful violations, (4) repeat violations, and (5) failure to abate violations (failure to correct a previous violation). Each of these categories carries the potential for substantial financial penalties assessed against organizations that fail to comply with applicable OSHA standards.

In addition to the potential for fines in the categories just explained, employers may also be penalized with additional citations and fines and/or prison if convicted of any of the following offenses: (1) falsifying records or any other information given to OSHA personnel, (2) failing to comply with OSHA's posting requirements, and (3) interfering in any way with OSHA compliance officers in the performance of their duties. In other words, regulatory compliance is an important issue that should be taken seriously by all personnel in an organization, from employees to the CEO. This message should be conveyed to higher management by safety professionals and the safety committee.

In my experience, the potential for financial penalties and prison sentences tactfully explained typically gets the attention of key decision makers in an organization. Consequently, when attempting to gain a commitment to a safety-first corporate culture, it is important to make sure that decision makers understand the potential for fines and prison. Such information should always be presented in a tactful, nonthreatening way. However, it should be part of the safety committee's rationale for a safety-first corporate culture, especially in cases where decision makers fail to appreciate the importance of workplace safety.

It is also a good idea for safety professionals to monitor their professional literature and make note of citations, penalties, and fines assessed against other organizations by OSHA and to circulate this information among decision makers. You might also want to remind decision makers in your organization that OSHA is empowered to conduct workplace inspections unannounced. Further, organizations that receive citations for high-gravity violations are subject to even stricter enforcement measures.

The nature of the penalties and the amounts of the fines will typically encourage decision makers in organizations to consider more seriously what safety professionals tell them about establishing and maintaining a safety-first corporate culture. Key decision makers who look at this type of information and think, "That could have been our organization," will be more open to doing their part to maintain a safe and healthy work environment.

The purpose of this step—gain commitment—in the model is to help safety professionals in an organization formulate a comprehensive but concise rationale for establishing and maintaining a safety-first corporate culture. This rationale can be used to gain the commitment of key decision makers in the organization for establishing and maintaining such a corporate culture. The effectiveness of all other steps in the model depends on gaining this commitment.

SAFETY-FIRST CORPORATE PROFILE
The Coca-Cola Company

The Coca-Cola Company is one of the most recognized organizations in the world, employing thousands of people in many different countries—a truly global corporation. Occupational safety and health are top priorities at Coca-Cola. The following quote comes from the company's statement of corporate social responsibility:

> At The Coca-Cola Company, our long-term success depends upon ensuring the safety of our workers, visitors to our operations, and the public. We are committed to conducting our business in ways that provide all personnel with a safe and healthy work environment.

To carry out this commitment, Coca-Cola developed and deployed a safety management system known as The Coca-Cola Safety Management System (TCCSMS). This safety management system incorporates quality, environment, safety and health, and loss prevention into one comprehensive framework. It is used in all of the company's operations worldwide. TCCSMS provides a rigorous set of operational and risk management controls, an internal audit process, external audits, orientation for new hires, and ongoing training for all personnel—all of which are among the types of strategies used by organizations that maintain a safety-first corporate culture. TCCSMS is evidence of the Coca-Cola Company's commitment to a safety-first corporate culture.

Source: http://www.thecoca-colacompany.com/citizenship/workplace_rights_policy.html

Persuasion Strategies for Safety Professionals

If you are one of the fortunate safety professionals in an organization that is already committed to a safety-first corporate culture, skip this section. Or, if your organization has recently experienced a safety, health, or environmental tragedy, executives, managers, and supervisors may have already decided to commit themselves to the establishment of a safety-first corporate culture. If this is the case in your organization, you too may skip this section. However, if you sense that executives, managers, and/or supervisors are not fully committed to a safety-first corporate culture, you and your colleagues on the safety committee have some work to do, the work of persuasion. If this is the case, the strategies in this section will help you.

Persuasion Strategies

Before presenting some strategies that will make it more likely for listeners to accept your proposal, there is an important point to be made. That point is simply this: When trying to persuade people of the value of a safety-first corporate culture, never forget that you are in the right. Do not feel that you need to apologize for what you will ask of decision makers in your organization. Remember, establishing and maintaining a safety-first corporate culture is the right thing for an organization to do from every point of view (i.e., competitiveness, compliance, cost, and corporate social responsibility). Committing to a safety-first corporate culture is good for executives, managers, supervisors, employees, customers, suppliers, and the community. In other words, it's good for business.

The following persuasion strategies can be used to make it easier for key personnel in your organization to accept the various recommendations you will make concerning establishing and maintaining a safety-first corporate culture.

Build Trust   An unalterable fact of life is that it is easier to persuade people who trust you than those who do not. Trust is an essential element in persuasion. This means that as you interact with personnel at all levels in your organization, do so in ways that build trust. Develop and nurture a reputation for telling the truth, even when it is unpleasant to do so. Be tactful when you have to deliver bad news, but be truthful. You do not protect people from bad news by downplaying or concealing it.

Follow through on promises you make. In other words, do what you promise to do, do it on time, and do it well. In fact, go beyond this commitment and adopt the following trust-building strategy: When dealing with people, promise small, but deliver big. In other words, never overpromise, and always do more than you promise. Finally, if you make a mistake, admit it, correct it, learn from it, and move on. Do not try to cover up mistakes or blame others for them. Few things will give you more credibility with other people than having the courage to admit that you have erred and to accept responsibility for the error.

Avoid Information Overload   Before trying to persuade anyone of anything, stop for a minute and ask yourself this question: How can I state my case in the shortest, most concise manner? Remember, decision makers in an organization are busy people who are constantly bombarded by information from an ever-growing list of sources. If you overload them with too much information, they will not be persuaded. In fact, information overload is more likely to generate opposition than support.

Avoid Condescending Tones and Technical Language   As a safety professional, you are going to know more about the safety-related recommendations decision makers must be persuaded to accept than they will, and you are going to speak the language of safety better than they do. It is easy when you know more than your audience to unconsciously slip into an attitude of condescension. Guard against this. Nobody likes to be talked down to. This is especially true of people who have achieved the status of key personnel in an organization. Avoid technical language others might not understand and condescending tones. Rather, speak to key personnel as professional colleagues and convert the language of safety into lay terms.

Listen Carefully to Their Concerns   When trying to persuade others of the validity of your point of view, it is easy to get so wrapped up in the points you are making that you fail to listen to the concerns raised by your audience. Persuasion requires two-way communication. Safety professionals who simply conduct one-way "broadcasts" rather than two-way conversations are not likely to receive a positive reception, especially from executives. In addition, by listening carefully to the concerns expressed by those you are trying to persuade, you can learn about (1) the areas of opposition; (2) the points of agreement; and (3) the need for reshaping, revising, or redirecting your message. When trying to persuade people of your point of view, look directly at them, listen carefully to what they say—verbally and nonverbally—and take mental notes that you can use later to reshape your message, if necessary.

Watch for Nonverbal Cues   In face-to-face conversations, much of the communication that takes place occurs nonverbally. This is especially true when you are trying to persuade others of the validity of your point of view. People that rise to the executive level in an organization are typically smart and articulate. Many, by necessity, have become adept at the art of verbal give-and-take, the kind of repartee that results in a skillfully worded noncommitment. To help sort out true commitment from skillfully worded, tactfully delivered noncommitment, safety professionals need to become adept at reading nonverbal cues.

Often the nonverbal cues people unconsciously provide during a conversation more accurately reflect their true feelings about what you are saying than their spoken words. Nonverbal cues come in the form of gestures, facial expressions, proximity, and rate of speech as well as voice tone, pitch, and volume. If this sounds complicated, relax. The good news is that you can read nonverbal communication right now, even without the benefit of special training. In fact, you could read nonverbal communication before you could speak or understand the first spoken word. A baby can tell if its mother is angry, sad, happy, or stressed by the nonverbal cues she provides (tenseness, tone of voice, volume of speech, etc.).

The keys to understanding nonverbal communication are (1) paying attention—listening with your ears and your eyes—and (2) watching for differences in what is said verbally and what is conveyed nonverbally, rather than for specific gestures. When you observe a difference between what is said verbally and what is conveyed nonverbally, there is a problem. For example, if an executive seems to be saying that he agrees with you, but avoids eye contact, do not assume you have his or her commitment. The words say "yes," but the nonverbal cue means "no."

I have experienced this situation several times. In one case, the CEO—speaking on behalf of the organization's executive management team—responded to my rationale for establishing a safety-first corporate culture by cutting me off and saying, "Yea, yea. I already know all of that. Everybody in this room agrees with you." This executive's words said he agreed, but his nonverbal cues conveyed just the opposite. The tone of his voice indicated irritation and boredom, and his hand gestures—as if waving me away—were dismissive. Clearly, I did not have a true commitment from this CEO.

Consequently, I decided to put his so-called commitment to the test. I did this by making several practical proposals such as revising the organization's strategic plan to include safety as a corporate value, developing a corporate safety policy, and revising job descriptions and performance appraisals to include safety. The executive almost fell out of his chair. By simply taking him at his word and giving him a few examples of how his so-called commitment would be translated into action, I was able to tactfully show this executive that a commitment to safety required more than just words. This approach had the salutary effect of leading to a productive dialogue concerning what his commitment would actually mean in practical terms, and, as a result, I was eventually able to win this executive's commitment.

When you find yourself in a situation where executives verbally commit to a safety-first corporate culture but their nonverbal cues create doubt in your mind, tactfully test your suspicions by recommending some of the practical actions covered in the next chapter and see what happens. This approach might lead to productive dialogue, or it might just show that you still have more persuading to do. In either case, it is best to know where you really stand before attempting to go forward.

Apply the Building Blocks of Persuasion   Building a persuasive argument in favor of your point of view is like building a house; it is best to begin by laying a strong foundation. What I call the "building blocks" for a persuasive argument are as follows: (1) facts, (2) interest/enthusiasm, (3) attitude, (4) flexibility, (5) tact, (6) courtesy, and (7) nonoffensive persistence.

- *Facts.* When trying to persuade people to your point of view, have the relevant facts, make sure they are accurate, and present them in a user-friendly format.

- *Interest/enthusiasm.* If you want others to be interested in what you have to say, show interest in it yourself and be enthusiastic. Proposing a safety-first corporate culture is a good thing—show your interest in it and be enthusiastic. If you aren't passionate about your ideas, why should anyone else be?

- *Attitude.* One of the most contagious of human phenomena is a positive attitude. People catch your attitude in the same way they catch your cold: by being around you. If you can maintain a positive attitude as you advocate for a safety-first corporate culture, eventually others will "catch" your positive attitude toward safety.

- *Flexibility.* One size rarely fits all, and there is usually more than one way to solve a problem. Consequently, as you make proposals relating to safety to key personnel, be flexible. The people whom you are trying to persuade might know something you don't or might see another way to do what you are proposing.

- *Tact. Tact* is sometimes referred to as making your point without making an enemy. Tact is critical when trying to persuade others to accept your point of view. You will not help your cause by making enemies of the people whose commitment to a safety-first corporate culture is needed.

- *Courtesy.* One of the most effective persuasion tools available to you is courtesy. Caustic, rude, and insensitive behavior will not win people over to your point of view. Just being courteous when trying to persuade others of the need for a safety-first corporate culture will not guarantee success, but failing to be courteous can ensure failure.

- *Nonoffensive Persistence.* People are rarely persuaded easily. It is probably unrealistic to expect that you will win key personnel over during your first meeting with them. Consequently, persistence will be critical. This does not mean you should become an irritant to the key personnel you are trying to persuade. Remember to be tactful and courteous, but don't give up. Keep pressing your point.

COLLEAGUE-TO-COLLEAGUE DISCUSSION CASES

CASE 1: The Unsupportive CEO

Tom Patterson is the new safety director at ABC, Inc., a manufacturing firm that employs more than 2,000 people. Patterson was brought in to help ABC, Inc., establish a safe and healthy workplace after the company received a substantial fine from OSHA for serious noncompliance issues. At first, Patterson thought he would have the support needed to get the job done, a task he described during an interview as "establishing a safety-first corporate culture at ABC, Inc." However, lately he is having trouble even getting time on the CEO's calendar. To make matters worse, since the CEO does not seem to be supporting his efforts, the other executives in the company are not either.

Patterson is beginning to feel like the CEO hired him, told him to fix "this mess," and then forgot about him. It's as if the CEO expects Patterson to establish a safety-first corporate culture without bothering him or the company's other executives. In Patterson's assessment, ABC, Inc., is one big accident waiting to happen. He needs to gain the support and commitment of the company's top executives if he is going to have any chance of changing the safety component of ABC's corporate culture. Unfortunately, Patterson feels stuck. He cannot get the job done without the support of ABC's senior managers, but he is at a loss as to how to gain their support and commitment.

Discussion Questions

1. Have you or any of your colleagues ever been in a similar situation? If so, describe the situation and how you handled it.

2. What advice would you or your colleagues give Tom Patterson concerning how to proceed in this situation?

CASE 2: Resistance from Supervisors

Mark Day has the support he needs from the CEO and executive managers to turn things around at XYZ Corporation. Right now, the company's safety record is the worst Day has ever seen. Accidents and injuries are so common that most employees hardly even react to them any more. In fact, in two years there have been only a handful of days in which there was no time lost due to an injury. Employee morale is about as low as it can get. If XYZ Corporation weren't the only employer in this one-horse town, the majority of the company's employees would leave. The fact that they can't just makes matters worse.

The dismal situation he now faces is one of the reasons Day agreed to accept his new position as safety manager for XYZ Corporation. OSHA fines, lawsuits, medical expenses, and declining productivity had gotten the attention of the company's top managers. Day could tell during his interview that the CEO and vice presidents were scared. If the situation isn't turned around, XYZ Corporation will be a thing of the past; its contracts sent to China or India. This is why Day is so surprised to be meeting such strong resistance to his efforts to establish a safety-first corporate culture at XYZ.

The resistance is coming from supervisors. They are the managers who are responsible for production rates. The company's supervisors know that Day has the support of higher management. Consequently, their resistance is covert. It amounts to removing safeguards from machines, pushing employees to neglect the use of personal protection equipment, and pressuring employees to do anything they can to maintain production rates, regardless of the hazards.

In safety meetings, the supervisors agree with Day to his face, but go right back to the shop floor and ignore his admonitions. Day is at a loss over how to handle this situation. He can go over the heads of supervisors to their bosses, but is afraid that if he does they will simply become even more covert in their resistance. Day needs the support and commitment of XYZ's supervisors, but is stumped regarding how to go get it.

Discussion Questions

1. Have you or any of your colleagues ever been in a similar situation? If so, describe the situation and how you handled it.

2. What advice would you or your colleagues give Mark Day concerning how to proceed in this situation?

Key Terms and Concepts

Before leaving this chapter, make sure you understand the following key terms and concepts and can accurately explain them to people who are not safety professionals.

Competitiveness rationale	Employee morale costs
Superior value	Corporate image costs
Peak performance	Ethics rationale
Continual improvement	Compliance rationale
Cost rationale	

Review Questions

Before leaving this chapter, make sure you can accurately and comprehensively, but succinctly, answer the following review questions:

1. Explain the competitiveness rationale for a safety-first corporate culture.
2. Explain the meaning of "superior value" and how it relates to establishing a safety-first corporate culture.
3. Explain how the concepts of "peak performance" and "continual improvement" relate to establishing a safety-first corporate culture.
4. Explain the cost rationale for a safety-first corporate culture.
5. Explain the ethics rationale for a safety-first corporate culture.
6. Explain the compliance rationale for a safety-first corporate culture.
7. List and explain the "building blocks" of persuasion.

Application Project

When trying to win the support and commitment of key decision makers in an organization for a safety-first corporate culture, remember, one size does NOT fit all. The approach adopted in one organization might not work in another. This is because the exigencies of every organization—although similar in many ways—can be different in other ways. Consequently, before beginning your efforts to win the support of decision makers in your organization, mentally assess the situation. Who is most likely to support you? Who is most likely to resist and why? On the basis of this assessment of the situation, you can develop a set of strategies for winning support and commitment that is tailored to your specific situation.

Your challenge in this project is to develop a tailored set of strategies for winning the support and commitment of executives, managers, supervisors, and employees for a safety-first corporate culture. After assessing the situation in your organization, proceed as follows:

1. Plan
2. Implement
3. Monitor
4. Adjust as necessary

Endnotes

1. European Agency for Safety and Health at Work. "Safe, Healthy, Competitive—the New Quality of Work Initiative in Germany," retrieved May 2008 from http://osha.europa.eu/publications/conference/20041201/index_5.htm

2. European Agency for Safety and Health at Work. "The Business Case of Safe and Healthy Work," retrieved May 2008 from http://osha.europa.eu/press_room/050121_CSR/newsarticle_view

3. Ibid.

4. From "Employee Perceptions: Impact of Work Factors on Job Performance," by the Institute for Corporate Competitiveness, Niceville, FL, Report 2003-11, June 2003.

Chapter 2

Expect Safety-First Attitudes and Practices

Major Topics

Safety expectations in
- The organization's strategic plan
- Corporate policy
- New-employee orientations
- Team charters
- Job descriptions
- Performance appraisals
- Everyday monitoring
- Role modeling
- Peer pressure
- Training requirements
- Mentoring
- Reinforcement

There is a management adage that says: "With most people, you get what you expect." This adage is typically used to admonish managers to aim high when setting performance goals for their direct reports. However, it may also be applied to workplace safety. The application is this: If you want safety-first attitudes and practices in your organization, it is important to expect such attitudes and practices and to show that they are expected.

If it is important to show that safety-first attitudes and practices are expected, safety professionals must understand how their organization can most effectively communicate such expectations to its personnel. There are many ways an organization can effectively communicate expectations to its personnel. These include (1) making safety expectations part of the organization's strategic plan—first as a corporate value and second as a competitive strategy, (2) developing and effectively communicating a corporate safety policy to all personnel, (3) including a comprehensive safety component in new-employee orientations, (4) incorporating safety as a ground rule in team charters, (5) including safety in all job descriptions, (6) including at least one safety criterion in all performance appraisal instruments, (7) monitoring safety-related attitudes and practices daily, (8) role modeling appropriate safety-related attitudes and practices, (9) ensuring peer pressure works in favor of safety, (10) providing ongoing safety training, (11) mentoring the appropriate safety-related attitudes and practices, and (12) reinforcing appropriate safety-related attitudes and practices using recognition and rewards.

Safety Expectations in the Organization's Strategic Plan

A well-developed strategic plan tells stakeholders what the organization is, why it exists, where it is going, how it plans to get there (competitive strategies), and what values will guide its operations and decisions (corporate values). More important than this, however, is the fact that an organization's strategic plan is an executive-level document that communicates what is strategically important to its executive management. As such, persuading executives in your organization to include safety in the strategic plan will test their commitment to workplace safety. This is another reason why it is so important that safety be included in the strategic plan as both a corporate value and a competitive strategy.

A comprehensive strategic plan should have at least the following elements: corporate vision, mission, and values; competitive strategies; and broad strategic goals. Figure 2.1 is an excerpt from a strategic plan that shows how and where safety imperatives fit into such a plan. Corporate values and competitive strategies that are linked directly or indirectly to safety appear in bold print.

Safety as a Corporate Value in the Strategic Plan

If an organization's mission statement explains why it exists—in other words, its purpose—corporate values provide the framework within which the mission will be pursued. There is an important message inherent in an organization's strategic plan—sometimes stated and sometimes not—that says: "As we pursue our mission, vision, and goals, we will be guided by the following corporate values." This is why it is so important to ensure that safety is included in the strategic plan as a fundamental corporate value.

The corporate values section of the partial strategic plan in Figure 2.1 contains three values that relate directly to safety. The first says: "We will provide our personnel a safe and healthy work environment that is conducive to peak performance and continual improvement." This statement makes clear the commitment of ABC Corporation's executives to workplace safety as well as their belief that a safe workplace promotes peak performance and continual improvement. This belief ties back into the company's vision that says: "ABC Corporation will be a global leader." To be a global leader in any field, an organization must perform at peak levels and continually improve all of its operations. ABC Corporation's executives clearly see a link between the company's ability to be a global leader and workplace safety.

The second safety-related corporate value in ABC Corporation's strategic plan says: "We will be good corporate neighbors and protect the environment from the hazardous

---

ABC CORPORATION

Mission Statement
ABC Corporation manufactures discrete metal components that are precision machined in small lots for our customers in the aircraft industry in more than 30 countries worldwide.

Vision Statement
ABC Corporation will be the global leader in the manufacture of precision-machined metal parts.

Corporate Values
As we pursue our mission and vision, ABC Corporation will be guided by the following corporate values:

- We will provide our customers with superior value (superior quality, cost, and service) in our products and services.
- We will provide our stockholders a favorable return on their investment.
- We will adhere to the highest ethical guidelines.
- *We will provide our personnel a safe and healthy work environment that is conducive to peak performance and continual improvement.*
- We will make customer satisfaction a top priority.
- *We will be good corporate neighbors and protect the environment from the hazardous by-products of our manufacturing processes.*
- We will treat all of our stakeholders—customers and suppliers—as partners.
- **We will treat our personnel as valuable assets to be appreciated, protected, nurtured, and developed.**
- We will continually improve our products, processes, services, and personnel.

Competitive Strategies
The following strategies will be used to maintain competitive advantage in the marketplace:

- *Our employee-positive values will help us to attract the most talented personnel available.*
- Our global marketing and sales partnerships will help us to maintain a steadily growing market share.
- *Our safe and healthy work environment will help us to retain the best personnel and free them to focus on peak performance and continual improvement.*
- Our Lean Sigma manufacturing practices will help us to produce world-class quality.

FIGURE 2.1   Strategic Plan (Excerpt)

by-products of our manufacturing processes." In today's "green-oriented" world, organizations must be environmentally friendly corporate neighbors in the communities in which they are located. To do less is to risk both image problems and costly litigation and fines.

The third safety-related corporate value in ABC Corporation's strategic plan says: "We will treat our personnel as valuable assets to be appreciated, protected, nurtured, and developed." This corporate value is not as directly tied to safety as the other two, but the link is still strong. "Valuable assets" that are to be "protected" strongly implies that the safety and health of employees is a high priority.

Safety as a Competitive Strategy in the Strategic Plan

Few things make so powerful a statement in an organization's strategic plan as the factors chosen as competitive strategies. An organization's competitive strategies are applied at the strategic level to set it apart from the competition. Competitive strategies are used to gain and maintain competitive advantage. Consequently, they must represent things in which the organization will invest resources in an attempt to do better than the competition, things done so well the competition will be unable to replicate them.

In Figure 2.1, ABC Corporation's strategic plan contains two competitive strategies that relate directly to safety. The first of these says: "Our employee-positive values will help us attract the most talented personnel available." As was explained in the previous section, three of these "employee-positive" values relate either directly or very closely to safety. An organization that can succeed in attracting the most talented personnel will, indeed, have a competitive advantage over competing organizations.

The second safety-related competitive strategy in Figure 2.1 says: "Our safe and healthy work environment will help us retain the best personnel and free them to focus on peak performance and continual improvement." Not only does ABC Corporation plan to attract the most talented personnel, it intends to retain them, secure the best possible performance from them, and continually improve their performance. The company's executives clearly view providing a safe and healthy work environment as a necessary step in achieving this desirable outcome.

As a safety professional, you can be sure you have the commitment of your organization's executives when they are willing to include safety in the strategic plan as a corporate value and a competitive advantage. A word of caution is in order here, though. Even when you have persuaded your organization's executives to include safety in the strategic plan, it may be necessary to show them how. There are two ways to achieve this. The first way is to persuade them to include your organization's chief safety professional on the strategic planning team. The other is for the safety committee to draft a set of safety-related corporate values and at least one competitive strategy and submit this material to the executive management team for inclusion in the strategic plan. A slightly different twist on this approach is for the safety committee to draft the material to be included in the strategic plan and ask the executive who serves on the safety committee to be responsible for getting it included in the plan.

Safety Expectations in Corporate Policy

The corporate values and competitive strategies contained in an organization's strategic plan must be put into action if they are to do any good. A critical step in doing so is to develop corporate policies to guide the actions of all personnel as they relate to the organizational imperatives in the strategic plan. If safety is a corporate value or a competitive strategy or both, the organization must develop and deploy a corresponding corporate safety policy. Figure 2.2 is an example of such a policy.

From this policy it is obvious that ABC Corporation is committed to the safety of its people, processes, and products. If your organization does not yet have a corporate safety policy, drafting one is an excellent project for the safety committee. If your organization

---

ABC CORPORATION

Safety Policy

ABC Corporation is committed to providing customers with products of superior value that are safe and exceed expectations. We are also committed to protecting the safety and health of our personnel, customers, suppliers, and the environment. To this end, it is the policy of ABC Corporation to

- provide a safe and healthy work environment for all personnel;
- comply with all applicable federal, state, and local safety regulations and guidelines;
- manufacture products that are safe for the use of customers;
- take appropriate steps to protect the environment from the hazardous by-products of our manufacturing processes; and
- take appropriate steps to conserve energy, water, and other natural resources.

FIGURE 2.2

has a policy, the safety committee should make sure it is sufficiently comprehensive to cover workplace, product, and environmental safety as well as regulatory compliance.

When safety has been included in the organization's strategic plan and a corresponding safety policy has been developed, the *big picture* requirements for establishing a safety-first corporate culture have been satisfied. The remaining requirements are more focused and more practical in nature. Satisfying these practical requirements will require you and the other members of the safety committee to work closely and persistently with human resources personnel, middle managers, supervisors, and employees. However, once you have sufficient executive-level commitment to get safety included in the organization's strategic plan and to deploy a corporate safety policy, you have cleared the most important hurdles in the process.

Safety Expectations in New-Employee Orientations

Orientations have always been important for new employees. An effective orientation makes new employees feel like part of the team, prepares them to succeed in their new job, and makes them aware of what is expected of them by the organization. For this reason it is important that your organization's employee orientation has a comprehensive safety component. New employees should come away from the orientation understanding that safety-first attitudes and practices are the norm in your organization. Integrating safety into your organization's employee orientation is the subject of Chapter 4.

Safety Expectations in Team Charters

If teamwork is a priority in your organization, safety-first attitudes and practices should be part of the team-building process. An effective way to make this happen is to include safety as a ground rule when developing or revising team charters. A team charter is a brief document that contains the team's name, its mission statement, and a set of ground rules.

The ground rules establish a framework within which all members of the team will operate as they work together to accomplish the team's mission. At least one of the ground rules should be about safety. Team charters are the subject of Chapter 7.

Safety Expectations in Job Descriptions

One of the most fundamental tools for showing employees what is expected of them is the job description. If safety-first attitudes and practices are expected of personnel in your organization, their job descriptions should say so. This means that the safety-related responsibilities of all personnel—executives, managers, supervisors, and employees—should be part of their respective job descriptions.

To ensure that all job descriptions contain the appropriate responsibilities relating to safety, you and the other members of the safety committee will need to work with your organization's human resources department. In fact, it is a good idea to have a member of your organization's human resources department—preferably its head—serve on the safety committee because much of what must be done following the recommendations in this and the remaining chapters of this book will require the cooperation of your organization's human resources department.

Job Descriptions for Executives

Job descriptions for executives should convey the message that they are responsible for setting the tone and providing the leadership to ensure that your organization has a safety-first corporate culture. They are to provide the broad organizational framework that will allow for the establishment and maintenance of such a culture. Here is a safety-related statement for the job description of an executive the safety committee can use as an example in developing a similar statement for your organization:

> *Individuals in this position are responsible for setting the tone and providing the leadership to ensure that our organization is able to maintain a safety-first corporate culture. This includes providing the direction, resources, and policies necessary to ensure that our organization has a safe and healthy work environment and is an environmentally friendly corporate neighbor.*

Once such a statement has been drafted, it should be submitted to the human resources department. This is one of the reasons it is wise to include a representative from that department on the safety committee. Having such a member will ensure a more positive reception for the work submitted to the human resources department by the safety committee. The chief human resources officer for your organization will submit the draft statement to the organization's executive management team for approval before revising job descriptions.

Job Descriptions for Managers

Job descriptions for middle managers should convey the message that they are responsible for setting a consistent example of positive safety-first attitudes and practices and for giving employees a voice on safety-related issues and concerns. In addition, they should factor safety considerations into the decision-making process and ensure that safety is an important criterion when giving recognition, rewards, promotions, wage increases, and bonuses. Finally, middle managers should understand from their job descriptions that they are responsible for carrying out and enforcing policies relating to safety as well as ensuring regulatory compliance.

What follows is a safety statement for the job description of a middle manager the safety committee can use as an example in drafting a similar statement for your organization:

> *Individuals in this position are responsible for helping our organization maintain a safety-first corporate culture. Responsibilities include but are not limited to the following: (1) consistently setting a positive example of safety-first attitudes and practices; (2) factoring safety considerations into the decision-making process; (3)making safety an important consideration in decisions about employee recognition, rewards, promotions, wage increases, and bonuses; (4) carrying out and enforcing polices relating to safety; (5) ensuring total compliance with all applicable local, state, and federal safety regulations; and (6) giving employees a voice concerning safety issues and concerns.*

As with the job description statement for executives, this statement should be submitted to the human resources department, which will, in turn, submit it to higher management for approval before including it in the job descriptions of middle managers.

Job Descriptions for Supervisors

Because supervisors are the face of management in the eyes of employees, they can be safety professionals' best allies or worst enemies. Which case becomes the reality safety professionals have to deal with depends on the attitudes of supervisors toward safety. Because they are the managers who work most closely with employees on a daily basis and because employees look to them for direction and guidance, supervisors must be on the team, so to speak, when it comes to safety-first attitudes and practices. Consequently, it is important to ensure that the job descriptions of supervisors contain a comprehensive statement about safety responsibilities.

Here is a job description statement for supervisors the safety committee can use as an example in drafting a similar statement for your organization:

> *Individuals in this position are responsible for ensuring that employees apply safety-first attitudes and practices as they do their jobs. To this end, supervisors are responsible for (1) involving employees in an ongoing program of hazard identification; (2) involving employees in developing and applying safe job procedures; (3) teaching employees how to properly use personal protective equipment and monitoring continually to ensure they do; (4) teaching employees good housekeeping practices and requiring their use; (5) teaching employees the fundamentals of safe work practices (i.e., safe lifting, safety glasses, etc.); (6) setting a positive example of safety-first attitudes and practices; (7) working with safety professionals, management, and the safety committee to ensure that our corporate safety policy is fully and effectively implemented on a daily basis and that our organization fully complies with all applicable safety regulations; and (8) working with safety professionals to investigate accidents, incidents, and near miss situations.*

This sample statement is long and detailed. If your organization considers it too long to include in the job descriptions of supervisors, just include item seven of the statement and provide the rest to supervisors as a separate checklist. As with the other job description statements drafted by the safety committee, this one should be forwarded to the human resources department for review and, in turn, to higher management for approval.

Job Descriptions for Employees

Employees represent the cutting edge of workplace safety. No matter how well executives, managers, and supervisors do their parts to establish and maintain a safety-first corporate culture, employees still determine, to a large extent, an organization's safety record. Employees must be willing to work safely themselves, expect their fellow employees to work safely, and assist management in identifying and correcting hazardous conditions.

Here is a job description statement the safety committee can use as an example in developing a similar statement for your organization:

> *Individuals in this position are expected to work safely, follow the organization's safety rules and regulations, and assist supervisors and managers in maintaining a safe and healthy work environment.*

As with other job descriptions developed by the safety committee, this one should be forwarded to the human resources department for review and, in turn, to higher management for approval.

Safety Expectations in Performance Appraisals

An effective way to show personnel what is expected of them is to make the expectations part of the performance appraisal process. Performance appraisals are used to improve performance. Consequently, if your organization wants to continually improve its safety-related performance, that performance should be evaluated. Further, all of the expectations set forth in job descriptions should have corresponding criteria in the performance appraisal instruments used by your organization. Evaluating safety-related attitudes and practices is covered in Chapter 8.

Safety Expectations in Everyday Monitoring

One of the most important ways supervisors and managers can show employees what is expected of them concerning safety-related attitudes and behaviors is to monitor those attitudes and behaviors on a daily basis. If safety is important, it should be monitored. If it is not monitored, employees will recognize that a disconnect exists between policy and practice. When a disconnect such as this occurs, employees will interpret it as follows: "Management does not mean what is says about safety. If it did, our supervisor would speak up when he sees us violating safety policies and practices." Monitoring safety-related attitudes and practices is covered in Chapter 8.

Safety Expectations in Role Modeling

Employees are more inclined to follow the examples of managers and supervisors than their words. Words are important, but they must be reinforced by the everyday actions of those who speak them. If a supervisor talks about the importance of safety-first attitudes and practices but fails to set a consistent positive example of them, his or her words will have no credibility with employees. Further, employees will follow the example rather than the words. This is the origin of the management adage that says: "Don't just talk the talk—walk the walk." Setting a consistent positive example of safety-first attitudes and practices is covered in Chapter 3.

Safety Expectations in Peer Pressure

Recall from Chapter 1 that the definition of a safety-first corporate culture includes a reference to the "unwritten rules" of the organization. In organizations that have a safety-first corporate culture, the unwritten rules support positive safety-related attitudes and practices. Unwritten rules are enforced primarily by peer pressure. Peer pressure originates in what people of equal status expect of each other, how people who have no formal authority to do so influence each other. Executives exert peer pressure on other executives. Managers exert peer pressure on other managers. Supervisors exert peer pressure on other supervisors. Finally, employees exert peer pressure on other employees. The key is to ensure that peer pressure is pro- rather than antisafety.

How does an organization ensure that peer pressure among its personnel works in support of safety? The big picture answer to this question is that organizations make peer pressure work on behalf of safety by consistently expecting, role modeling, orienting, mentoring, training, team building, monitoring, evaluating, reinforcing, and assessing for safety. In other words, organizations can ensure that peer pressure reinforces safety-first attitudes and practices by effectively implementing the 10-step model explained in this book. If employees are expected to apply prosafety peer pressure to other employees, they must see this happening at other levels in the organization. Further, they must know they have the backing of their supervisor, management, and the organization. The following example from my experience shows how supervisors can promote prosafety peer pressure and how they can undermine it.

A Tale of Two Supervisors

Although he was still a new employee at a company I will call Technical Associates, Inc. (TAI), Pete knew the company stressed safe work practices. Safety had been a major component of the orientation he and other new employees were required to complete during their first week on the job. Safety was a requirement in his job description, and he had completed a mandatory two-day safety training program just the previous week. Consequently, he was shocked when his supervisor took him aside and chewed him out for encouraging another employee to reinstall the machine guard he had removed from his stamping machine.

Pete and his colleague operated identical stamping machines on the day shift for TAI. Facing a difficult production deadline, Pete's colleague, Sam, removed the machine guard on his stamping machine. This allowed him to feed the material into the machine faster, but it also created a hazardous condition in which just one mental lapse on Sam's part could cause a serious injury. Knowing that TAI expected safe work practices, Pete questioned Sam and told him he should reinstall the machine guard. Sam told Pete to mind his own business. Pete responded by saying, "We won't make our deadline if you lose a hand in that machine."

Hearing this exchange, their supervisor took Pete aside and told him to stop interrupting Sam and to focus on getting his own work done on time. Pete was shocked. He was sure he had read the following line in one of the documents he had been given during his orientation: "At TAI we want safety and productivity, but we will settle questions between them in favor of safety. We will not attempt to increase productivity through the use of unsafe work practices."

In another case, John was a forklift driver for a company I will call Manufacturing Associates, Inc. He had a reputation for being the company's best driver. John's productivity numbers were consistently 8 to 10 points higher than those of his fellow drivers. Unfortunately, John achieved a better productivity rating than his colleagues by ignoring important safety rules.

One day, after a near-miss incident in which John came close to running over another employee, his fellow drivers cornered him and told him they had had enough. One of them said, "If you have to kill someone to exceed your productivity quota, you need to think about your priorities." John responded by telling his fellow drivers they were just jealous because he consistently out-produced them. Then he laid down a challenge: "If you don't like it, bring the issue up in our next team meeting, and let's see what our supervisor has to say."

At the next team meeting, John's fellow drivers called his bluff and raised the issue. After some heated discussion in which John said his higher productivity numbers justified his penchant for cutting corners, the supervisor shocked him by siding with the other drivers. He pulled an article from a safety newsletter out of his pocket and read it to the team of drivers. The article talked about a tragedy that had occurred in which two people were killed and several others seriously injured when a forklift sped around a blind corner and slammed into a group of employees who were all bent over a table reading a set of blueprints.

According to the article, the company had been assessed a heavy fine by OSHA, the driver and his supervisor had been fired, and the company—along with the driver and his supervisor—had been sued for damages in civil court. John's supervisor told all of the drivers, but he looked squarely at John when he said it, that he expected them to make their productivity numbers and was pleased when they exceeded them. However, they were not to make their numbers by ignoring the company's safety rules. Then he put some teeth into his admonition on behalf of safety when he said, "From this point forward, any driver caught disobeying our safety rules will be recommended for immediate termination."

In the first story, the lesson Pete learned was that the company's emphasis on safety was all talk—at least in his team. He had tried to apply prosafety peer pressure with a fellow machine operator—just as he had been instructed to do during his orientation—but quickly found that his supervisor would not support him. To Pete, if the supervisor would not support him in the application of prosafety peer pressure, the company did not really believe in safety. His job description, orientation, and training had been nothing but words.

In the second story, John had assumed that productivity would trump safety. He was sure that his consistently high productivity would protect him from the consequences of breaking the company's safety rules. When his fellow drivers put this assumption to test, John quickly learned how wrong he was. Not only did productivity not trump safety, failing to follow established safety rules could get him fired regardless of his productivity rating. Because of how the supervisor in this case handled the situation, John's fellow drivers learned that prosafety peer pressure did work and did have the company's support.

These two stories make the critical point that if your organization expects peer pressure to work in favor of safety-first attitudes and practices, employees have to see hard evidence that using peer pressure to encourage workplace safety is the right thing to do. That evidence must come in the form of consistent examples of the organization's commitment to a safety-first corporate culture.

Safety Expectations in Training Requirements

If your organization thinks certain attitudes and practices are important, it follows that training relating to these attitudes and practices will not just be provided, but will be required. Employees who are required to complete ongoing safety training will get the message that safety is important. Providing ongoing training relating to safety-first attitudes and practices is the subject of Chapter 6.

Safety Expectations in Mentoring

The most competitive organizations provide new employees with mentors to help them gain a sufficient foothold to succeed in their new positions. Organizations that include safety-first attitudes and practices in what mentors are required to teach new employees send a powerful message about the importance of safety and expectations relating to it. Mentoring for safety-first attitudes and practices is the subject of Chapter 5.

Safety Expectations in Reinforcement

Competitive organizations make a point of reinforcing the attitudes and practices they deem desirable in their personnel. In fact, there is a management maxim that says: *Be sure to reinforce the attitudes, practices, behaviors, and performance you deem critical to success.* One of the best ways to reinforce safety-first attitudes and practices is by recognizing and rewarding the personnel who consistently exhibit them. This is the subject of Chapter 9.

SAFETY-FIRST CORPORATE PROFILE
Alcan, Inc.

Alcan, Inc., is a global leader in such operations as bauxite mining, alumina processing, metal smelting, power generation, aluminum fabrication, and specialty packaging. The company employs more than 65,000 people in 61 countries and regions. Mining, processing, power generation, fabrication, and packaging are all potentially hazardous operations. Add to this the size and geographical diversity of Alcan, and it is easy to see that maintaining a safety-first corporate culture is a challenge. However, it is a challenge the company meets.

Alcan maintains a comprehensive, systematic safety, health, and environmental management program. According to the National Safety Council (NSC), the company's program has resulted in more than $43 million in savings "while fostering a mindset and culture of EHS believers and champions throughout the company worldwide." The substantial savings that result from its commitment to workplace and environmental safety allow Alcan to reinvest in continual improvement initiatives that make it even more competitive.

Alcan began building a safety-first corporate culture by establishing clear expectations relating to safety. These expectations are summarized in the company's safety and health vision: "To be a recognized leader of safety, health, and environmental excellence in everything we do and everywhere we operate." All of Alcan's safety, health, and environmental excellence policies, programs, and practices grow out of this broad vision.

Source: http://www.nsc.org/news/nr110606.htm

COLLEAGUE-TO-COLLEAGUE DISCUSSION CASES

CASE 1: Uncooperative Supervisors

Janice Pilcher is the director of safety, health, and environmental management at QRS Corporation—a medium-sized defense contractor. Several months ago, she and her colleagues on the company's safety committee undertook an organization-wide effort to establish a safety-first corporate culture at QRS Corporation. Pilcher is satisfied she has the support and commitment of higher management, but she thinks the safety committee's efforts are being undermined by the company's supervisors. The problem appears to be productivity versus safety.

As she and other members of the safety committee have looked into the situation, they have come to the conclusion that employees are getting mixed messages concerning what is expected of them. Higher management is cooperating in setting the tone and providing the resources and support to establish a safety-first corporate culture at QRS, but the safety-first message appears to be stopping at the level of the supervisors.

The supervisors Pilcher and her colleagues on the safety committee have interviewed say all of the right things about safety, but the employees interviewed tell a vastly different story. Supervisors whose teams fall behind on their production schedules appear to be telling employees to take dangerous shortcuts. Hurry-up tactics such as removing machine safeguards and failing to use appropriate personal protective equipment appear to be common on the shop floor. Pilcher and her colleagues on the safety committee plan to meet for the sole purpose of discussing how they can (1) show supervisors what is expected of them, (2) gain a real commitment to safety from supervisors, and (2) convince supervisors to effectively communicate the message of safety-first expectations to their direct reports.

Discussion Questions

1. Have you or any of your colleagues ever been in a situation in which supervisors undermined their organization's safety efforts? If so, describe the situation and how you handled it.

2. What advice would you and your colleagues give Janice Pilcher about how to handle this situation?

CASE 2: No Expectations of Safety in This Company

Marvin Brackin is facing the most difficult challenge of his career in occupational safety. He has been hired as a consultant to help LMN, Inc., recover from a major incident in which two employees were killed and several others injured. The investigations are still being conducted, and OSHA is dealing with the company's CEO concerning fines and other issues, but these are not Brackin's concerns. Regardless of what happens with OSHA, the company's senior managers now understand that things have to change at LMN, Inc., in the area of workplace safety. This is why they have contracted with Brackin.

Brackin is developing a plan for establishing a safety-first corporate culture at LMN, Inc. He has begun some preliminary work and is at the point where he needs to decide how to make sure that all personnel in the organization have the appropriate expectations relating to safety and how to get those expectations communicated effectively throughout the organization. This will be a critical component in his plan because, as Brackin has learned from observations and interviews, there are no expectations of positive safety-related attitudes and practices in the company.

Discussion Questions

1. Have you or any of your colleagues ever worked in an organization that had no expectations of positive safety-related attitudes and practices? If so, talk about the problems this situation caused.

2. What advice would you and your colleagues give Marvin Brackin concerning how to proceed with establishing appropriate safety-first expectations throughout LMN, Inc.?

Key Terms and Concepts

Before leaving this chapter, make sure you understand the following key terms and concepts and can accurately describe them to people who are not safety professionals.

Safety expectations in
the strategic plan

corporate policy

new-employee orientations

team charters

job descriptions

performance appraisals

everyday monitoring

role modeling

peer pressure

training requirements

mentoring

reinforcement

Review Questions

Before leaving this chapter, make sure you can accurately and comprehensively, but succinctly, answer the following review questions:

1. Explain how to use the organization's strategic plan to communicate safety-first expectations.
2. Explain how to use corporate policy to communicate safety-first expectations.
3. Explain how job descriptions can be used to communicate safety-first expectations to executives, managers, supervisors, and employees in your organization.
4. Explain how your organization's performance appraisal process can be used to communicate safety-first expectations to executives, managers, supervisors, and employees.
5. Explain how an organization can make sure that peer pressure works in support of safety-first attitudes and practices.
6. List the other methods that can be used for communicating safety-first attitudes and practices that were mentioned in this chapter but are covered in detail in later chapters.

Application Project

For this application project, you need to develop the following: (1) a core value statement relating to safety for your organization's strategic plan; (2) a competitive strategy statement relating to safety for your organization's strategic plan; (3) safety-related job description statements for executives, managers, supervisors, and employees in your organization; and (4) safety-related performance appraisal criteria for executives, managers, supervisors, and employees in your organization.

Chapter 3

Role Modeling the Expected Safety-First Attitudes and Practices

Major Topics

Role modeling
- Safety-related expectations
- Safety-first practices
- Characteristics that promote safety-first attitudes

Effective leadership is critical to the success of any organization trying to achieve peak performance and continual improvement—two goals that require a safe and healthy work environment. Few aspects of leadership could be more important in the current context than setting a consistent, positive example for employees. All personnel in positions of authority in your organization must be positive role models of the safety-first attitudes and practices they expect of employees. They must be seen at all times—on good days and bad—exhibiting the safety-related characteristics your organization is trying to instill in employees.

The examples of those who lead organizations and the functional units within them are just as contagious as the common cold. Set a positive example, and employees will "catch" it by just being around you. Set a negative example, and employees will be "infected" by it, again by just being around you. This is an important message for safety professionals to convey to executives, managers, and supervisors in their organizations.

Role Modeling Safety-Related Expectations

Role modeling your organization's safety-related expectations is one of the most important responsibilities mentors, supervisors, managers, and executives have when trying to instill proper attitudes and encourage proper practices in employees. It cannot be overemphasized that employees must see people in leadership positions *walking the walk* rather than just *talking the talk* when it comes to safety-related expectations. Executives, managers, and supervisors who expect certain attitudes and practices from employees must be the personification of those expectations at all times, and not just when they feel like it.

Role Modeling Safety-First Practices

Role modeling expectations relating to safe work practices involves setting a positive example of complying with all applicable safety rules and regulations in such areas as the following:

- Ergonomics
- Mechanical hazards and machine safeguarding
- Falling and impact hazards
- Lifting hazards
- Vision hazards
- Temperature hazards
- Pressure hazards
- Electrical hazards
- Fire and life-safety hazards
- Industrial hygiene hazards
- Confined space hazards
- Radiation hazards
- Noise and vibration hazards
- Automation-related hazards
- Blood-borne pathogen hazards

In all of these areas of concern, there are applicable OSHA standards and, on occasion, additional local rules, regulations, and procedures. Employees must see the organization's leadership personnel complying with all applicable standards, rules, regulations, and procedures in these areas and any others relating to safety. For example, it does no good to require employees to complete mandatory ladder safety training only to have them see managers and supervisors ignore the rules by walking under ladders, something I have seen numerous times when working as a consultant. As another example, it does no good to require employees to complete safe lifting training only to have them see managers and supervisors ignoring proper procedures when lifting heavy objects.

I once served as a consultant for a company that had experienced a significant increase in its workers' compensation costs as a result of an upsurge in accidents requiring medical treatment. The CEO of the company was convinced that workers' compensation fraud was running rampant in his company. My job as a consultant was to observe for a few days and make recommendations.

On my first day, the CEO took me for a tour of the shop floor. As we entered the shop area, there was a rack of safety glasses and hardhats available for visitors to the floor. He ignored them, in spite of the fact that I put on both. As we toured, the CEO strode right through an area that had been blocked off with warning cones and, in the process, stepped in a puddle of spilled oil he then spread across the floor as we continued the tour. Already concerned about this CEO's bad example, I got an even bigger shock—as did the employee in question—when he walked up behind a machine operator and, by way of a greeting, gave him a hearty but unexpected slap on the back that almost knocked him into the cutting tool of his machine. It soon became apparent that this CEO was a walking disaster area— a terrible example for his employees.

What I observed during my brief three-day visit to this company was managers, supervisors, and employees who consistently followed their CEO's bad example. The necessary rules and regulations were in place, but everyone simply ignored them. Small wonder the company's workers' compensation costs were increasing. For my exit conference with the CEO, I brought along a small pocket mirror. When he asked me what I thought the main problem was, I simply held up the mirror so he could see his reflection, a tactic that won no prizes for customer service, but at least conveyed an accurate and important message.

Role Modeling Characteristics That Promote Safety-First Attitudes

Helping executives, managers, and supervisors understand that they must be role models of safe work practices is important and, typically, not too difficult to do. However, when it comes to helping them understand that they must also be exemplary role models of certain characteristics that promote safety-first attitudes, the task can be more challenging.

Nevertheless, executives, managers, and supervisors must understand that in addition to consistently setting a positive example of following safety procedures, it is also important to set a positive example of displaying certain characteristics that promote safety-first attitudes. Safety professionals need to (1) understand how the characteristics in question can affect the attitudes of employees toward safety and (2) help executives, managers, and supervisors develop this same understanding. Helping executives, managers, and supervisors appreciate how their attitudes toward safety affect those of employees is a challenge for safety professionals and the safety committee.

The characteristics that can promote safety-first attitudes in employees include the following:

- Integrity
- Initiative
- Positive change agency
- Critical thinking
- Effective communication
- Positive attitude
- Acceptance of responsibility/accountability

Integrity and Safety-First Attitudes

The integrity of your organization's leaders can affect the attitudes of employees toward safety. One of the complaints often heard about today's generation of employees—a generation that has been called, among other things, the "me-generation" and the "entitlement generation"—is that they have little concern for rules and regulations. This generation of employees—the 18- to 30-year-old population cohort—grew up during an era in which cheating was widely accepted and rules were optional, a contention well documented by David Callahan in his book *The Cheating Culture.*

When individuals of the me-generation begin their careers, organizations such as yours need them to follow the rules and apply peer pressure to their fellow employees in ways that encourage them to follow the rules. Transforming the attitudes of me-generation employees toward rules and regulations is a difficult enough challenge for organizations under even the best of circumstances. It simply cannot be done if leadership personnel in your organization are observed being less than exemplary role models of integrity, especially as it relates to workplace and environmental safety.

For example, executives who approve including safety as a core value in their organization's strategic plan and then refuse to provide the resources necessary to maintain a safe work environment set a bad example that sends the wrong message to employees. Managers who require employees to complete safety training but then casually ignore the rules about wearing hardhats and safety glasses on the shop floor set a bad example that sends the wrong message. Supervisors who post safety rules and then look the other way when employees ignore them set a bad example that sends the wrong message. In each of these situations, the leader in question is displaying a flagrant lack of integrity, a fact employees will readily notice and a bad example they are likely to emulate.

Initiative and Safety-First Attitudes

In organizations that establish a safety-first corporate culture, employees are encouraged to take the initiative in identifying hazardous conditions and bringing them to the attention of management. In fact, safety-first organizations do more than just encourage employees to take the initiative in identifying hazardous conditions; they take the next step and establish systems and procedures that make it simple and convenient for employees to do so. They also empower them to recommend solutions for eliminating or mitigating the hazards.

It has been my experience over many years that organizations with poor safety records tend to hire employees *from the neck down.* In other words, they want employees to work, not think, to do only what they are told, rather than taking the initiative to propose better ways. High-performing organizations with safety-first corporate cultures have learned the fallacy of this approach. In today's hypercompetitive business environment, organizations need their employees to take the initiative in identifying ways to solve problems, eliminate hazards, continually improve performance, and enhance the quality of the work environment. In other words, they need them to think. This is especially important in relation to safety.

Isbell Construction Company makes sure that managers, supervisors, and employees understand that they are to take the initiative in identifying hazardous conditions and making recommendations for eliminating or mitigating them by putting this expectation in writing.

Isbell has a comprehensive safety policy that explains in detail what managers, supervisors, and employees are responsible for in relation to workplace safety. For example, supervisors are required, among other things, to hold safety meetings in which they are to involve their direct reports in continually improving the work environment. Employees are required, among other things, to "participate in safety committee meetings, training sessions, and surveys as requested and provide input into how to improve safety."[1]

It is the last part of this statement that is important in the current context. Employees are required to "provide input into how to improve safety." This means that not only does Isbell Construction want its employees to take the initiative regarding safety, it requires them to and puts the requirement in writing. Further, the company requires managers and supervisors to carry out the various provisions of its safety policy.

Change Agency and Safety-First Attitudes

Everything relating to workplace safety is in a constant state of change. OSHA's safety standards, state safety regulations, organizational procedures, personal protective equipment, safety technologies, and everything else relating to safety are updated and improved continually. In addition, continual improvement is an unalterable requirement for success in the global marketplace, and continual improvement means continual change.

If change in organizations is inevitable, it follows that in order to maintain a safe and healthy work environment, organizations and the people in them must be willing to change on a regular basis. In fact, they must be willing to embrace change as a normal part of their jobs, rather than as something they must reluctantly go along with. People and organizations that lull themselves into a state of complacency with the status quo soon fall behind in the marketplace.

Consequently, it is important that leaders in organizations be observed doing their part to help organizations anticipate and respond effectively to change. Role modeling strategies for executives, managers, and supervisors relating to change agency include (1) making sure all personnel understand the need for change and the role they play in making it happen, (2) anticipating the need for change, (3) gaining the commitment of personnel to specific change initiatives, (4) empowering employees to be positive change agents, and (5) and playing a positive role in executing changes.

During one of the interviews conducted during the development of this book, I met a manager who was especially good at gaining the commitment of his direct reports when changes needed to be made. Because of his excellent role modeling, his department had a reputation for being on the cutting edge of safety, quality, and productivity improvements. This manager—I will call him Pablo—had been both a safety manager and a quality manager during his career. As a result, he was an expert at continually improving safety, processes, quality, the work environment, and the performance of employees—a fact that accounted for his elevation to vice president of production for a medium-sized technology company of 900 employees.

One of the things that made Pablo especially effective is that he understood the importance of gaining employee commitment to proposed changes. Further, he knew how to go about winning that commitment. Having spent time as both a safety and a quality manager before becoming his company's production vice president, Pablo had seen his share of change initiatives crash and burn—many of them excellent initiatives—because uncommitted employees who opposed them either sat back and let them fail or, worse yet, actively sought to make them fail. Consequently, Pablo never undertook a change initiative without first gaining a sufficient level of commitment from the employees who would have to execute the change as well as from other stakeholders who would be affected by it.

Several people interviewed during the development of this book told me of organizational changes that had been "inflicted" on them by higher management. They described feeling surprised, left out, in the dark, and resentful concerning changes that management, as one employee described it, "sprung on us from out of nowhere." On the other hand, a few people told of supervisors like Pablo, who worked with them in making change rather than just telling them it had to happen and expecting them to go along with it.

What these change-positive supervisors had in common with Pablo is that they (1) made helping their organization anticipate change part of what was expected of their team members; (2) welcomed improvement recommendations from team members; (3) kept team members well informed concerning any changes higher management was considering; (4) used team meetings to allow team members to voice their opinions, suggestions, and frustrations concerning planned changes; (5) involved the team in planning for changes that had to be made; and (6) approached change from the perspective of "we" rather than management against employees.

This is the approach to change that you and your colleagues on the safety committee should advocate with managers and supervisors. In fact, one of the most important training sessions that can be included in your organization's overall training program is "change management." Managers and supervisors need to learn the six-step approach to managing change set forth in the previous paragraph, and they need to apply it to ensure that safety-related changes are effectively implemented in their organization.

SAFETY-FIRST FACT

The Safest Facilities Are the Most Productive

According to Ironore Company of Canada, the safest facilities are the most productive facilities. "Creating a culture in which everyone contributes responsibly and proactively to their own safety and the safety of their co-workers requires many things to come together: awareness, information, and skills, as well as the safe installation, operation, and maintenance of facilities and equipment. It also requires making safety a value even higher than that of such key areas as quality, productivity, and cost efficiency. In fact, we believe that the safest facilities are the most productive, have the lowest costs, and consistently deliver the best practices."

Source: Retrieved from http://www.ironore.ca/main/index.php?sec=4&loc=44&lng=EN on January 31, 2008.

Critical Thinking and Safety-First Attitudes

Earlier in this chapter, I related the example of the CEO who attributed a sharp rise in the cost of his organization's workers' compensation insurance to rampant workers' compensation fraud. You will recall that in this case the CEO, rather than workers' compensation fraud, was the real culprit. When it came to following safety procedures, he was such a bad example that his employees were simply following his lead and ignoring them. Hence, the higher accident and injury rates that were, in turn, driving up the cost of his organization's workers' compensation insurance.

This CEO was not a critical thinker, at least not when it came to workplace safety. Rather than collect the facts and consider them objectively—hallmarks of a critical thinker—he made assumptions about a serious problem, assumptions that turned out to be false. Hazard identification and analysis are important elements of the overall safety program of organizations that maintain a safety-first corporate culture. These elements require objectivity, critical thinking, and logic. If employees are going to be expected to approach hazard identification and analysis objectively, critically, and logically, they must see executives, managers, and supervisors do so.

Communication and Safety-First Attitudes

Executives, managers, and supervisors in your organization have probably become accustomed to the benefits of the so-called age of communication. They are probably accustomed to 24-hour news programs that air seven days a week, instant communication via the Internet, and almost constant interaction via cell phones. Consequently, on the one hand, they are accustomed to instant information and constant human interaction—all aided by technology—but on the other hand, many, as a result of technology, have developed a short attention span.

People with short attention spans are typically not good listeners—an essential component of effective communication—and few things are more important to promoting safety-first attitudes among employees than management personnel who will listen. In fact, role modeling effective listening is an essential requirement of management personnel in organizations that maintain a safety-first corporate culture. If management personnel in your organization expect employees to participate in maintaining a safe and healthy workplace, they must listen to what employees have to say. Said another way, they must be good role models of effective listening.

In addition to listening patiently and assertively to employees concerning safety-related issues and concerns, executives, managers, and supervisors must also keep them well informed. This means making sure that employees always have the latest information needed to do their jobs safely and to do their part in helping maintain a safe working environment. Including a representative number of employees on your organization's safety committee, sharing safety information during team meetings, posting information on bulletin boards as appropriate, having a dedicated safety section in your organization's newsletter, producing a monthly safety bulletin, and sending regular safety updates by e-mail to employees are all ways of role modeling effective communication as it relates to safety.

Positive Safety-First Attitudes

Developing positive attitudes toward safety or anything else in personnel can be a difficult challenge, but it is a challenge that organizations must meet if they are going to be competitive. A rule of thumb that I share with executives, managers, and supervisors all the time is this: *Don't let your negative attitude about safety give employees an excuse for developing a bad attitude.* When trying to instill safety-first attitudes in employees, positive examples are critical. Consequently, consistent role modeling of safety-first attitudes is a must for leaders in your organization.

I mentioned earlier that people will catch your attitude in the same way they catch your cold: by just being around you. If leaders in your organization consistently display a positive attitude toward safety, employees will "catch" a positive attitude. Correspondingly, if organizational leaders display a negative attitude toward safety—whether it is about OSHA regulations, local safety procedures, workers' compensation issues, or any other safety-related topic—employees will catch a bad attitude.

Earlier in this chapter, Isbell Construction was cited as an example of an organization that went to great lengths to communicate what is expected of managers, supervisors, and employees in relation to workplace safety. Isbell puts these expectations in writing as part of its corporate safety policy. One of the company's highest expectations is that all personnel will maintain and display positive attitudes toward safety. In Section C of the company's safety policy, under the heading "Assignment of Responsibilities," Isbell summarizes its expectations concerning safety-first attitudes: "Everyone should have a safe attitude and practice safe behavior at all times."[2] This might be the most important sentence in what is a comprehensive, well-prepared safety policy.

Responsibility/Accountability and Safety-First Attitudes

Even in companies that maintain a safety-first corporate culture, the ultimate challenge is to get personnel at all levels to buy into the concept that safety is everyone's responsibility, to accept that they are responsible for working safely and should be held accountable for doing so. This is more likely to happen in organizations where executives, managers, and supervisors are good role models of taking responsibility and accepting accountability for safety.

Setting a positive example of responsibility and accountability is a two-step process. The first step involves doing those foundational things set forth in this book such as having a comprehensive safety policy; including safety in the organization's strategic plan, job descriptions, performance appraisals, and team charters; making safety part of the organization's orientation program; requiring ongoing safety training of all personnel; providing safety-first mentors; monitoring safety-related attitudes and practices regularly; and so on. Executives, managers, and supervisors who truly accept responsibility for safety in their organizations will do these things.

Once the foundational steps just listed have been accomplished, executives, managers, and supervisors must then be seen setting a consistent positive example of living up to them. For example, the safety policy for Isbell Construction specifies that managers are responsible for—among other things—providing "sufficient staffing, funds, time, and equipment so that employees can work safely and efficiently."[3] If the organizational leaders at Isbell Construction do not set the right example by providing the specified resources, employees will think the company's safety policy is just words and will ignore it. However, if they see the company's leaders doing what must be done to provide the necessary resources in good times and bad, they will understand that they too must take responsibility for doing their part to maintain a safe and healthy workplace.

COLLEAGUE-TO-COLLEAGUE DISCUSSION CASES

CASE 1: The Quality Supervisor Is a Bad Role Model

During the monthly safety committee meeting, a supervisor who represents production personnel complained to another supervisor. His exact words were as follows: "The quality supervisor is a bad role model." The safety committee's chairman asked him to explain what he was talking about. It turned out that the supervisor of the quality management department was setting a bad example for employees as he made his rounds checking on the progress of various quality initiatives.

The quality supervisor rarely wears a hardhat or safety goggles in the shop area, never wears ear plugs—even in areas with uncomfortably high noise levels—and is frequently seen taking shortcuts through areas marked off as hazardous. In short, the supervisor in question simply ignores safety procedures when he thinks they are inconvenient. Several of the employees on the production floor have mentioned the quality supervisor's bad example to their supervisor. The production supervisor has talked with his colleague from quality management, but so far nothing has changed.

The production supervisor's reason for bringing the issue up during the safety committee meeting is to solicit help. He has noticed that several production employees are beginning to follow the quality supervisor's bad example and wants advice concerning how he should handle this situation. On the one hand he wants the quality supervisor to cooperate in being a positive example of following safety procedures, but on the other hand, he has no line authority over him.

Discussion Questions

1. Have you or your colleagues ever been in a similar situation? If so, describe the situation and how you handled it.

2. What advice would you or your colleagues give the production supervisor concerning how to handle this situation?

CASE 2: Worst Safety Record in the Company

The shipping and receiving department for ABC, Inc., has the worst safety record in the company. Lifting, falling, and impact injuries are common in this department. As the director of safety for ABC, Inc., you plan to undertake an investigation to determine why the accident, incident, and injury rates in this department are so high. After all, shipping and receiving employees receive the same safety-related orientation, training, and mentoring that other employees do.

Your first interview is with the shipping and receiving department's supervisor. Within 10 minutes the source of the safety problems in this department is obvious to you. The supervisor's attitude toward safety is unabashedly negative. Several times she has referred to safety as a "necessary evil." Further, she takes no responsibility for her department's poor safety record and does not see why she should be held accountable for it. In her words, "After all, you are the safety director. Safety is your problem."

Discussion Questions

1. Have you or your colleagues ever had to deal with a supervisor who had a negative attitude toward safety and took no responsibility for it? If so, describe the situation and how you handled it.

2. What advice would you or your colleagues give the safety director in this situation concerning how to handle it?

Key Terms and Concepts

Integrity and safety-first attitudes

Initiative and safety-first attitudes

Change agency and safety-first attitudes

Critical thinking and safety-first attitudes

Communication and safety-first attitudes

Positive management attitudes and safety-first attitudes

Responsibility/accountability and safety-first attitudes

Review Questions

1. Explain why it is important that employees see people in leadership positions walking the walk rather than just talking the talk when it comes to safety-related expectations.

2. Explain how executives, managers, and supervisors can go about role modeling safety-first practices.

3. Explain how organizational leaders can role model integrity as it relates to safety.

4. Explain how organizational leaders can role model initiative as it relates to safety.

5. Explain how organizational leaders can role model positive change agency as it relates to safety.

6. Explain how organizational leaders can role model critical thinking as it relates to safety.

7. Explain how organizational leaders can role model effective communication as it relates to safety.

8. Explain how organizational leaders can role model positive attitudes as they relate to safety.

9. Explain how organizational leaders can role model accepting responsibility and accountability as they relate to safety.

Application Project

There is a maxim that says: "It is easier to state your beliefs than to follow them." This maxim applies directly to organizational leaders who should be good role models of the attitudes and practices they expect of employees. It is one thing for organizational leaders to formally communicate safety-related expectations to employees through such means as a comprehensive safety policy, and it is quite another for them to be good role models of these expectations.

To ensure that key personnel in your organization are good role models of safety-related expectations, give the safety committee the following assignment:

1. Each member is to observe executives, managers, and supervisors as they interact with him or her on a daily basis for one month. Make notes concerning instances in which they are good and bad role models, respectively.

2. Bring the notes on good and bad examples to a designated safety committee meeting. Were there more good or bad examples? Does your organization have a problem with regard to the safety-related examples of its leaders, or are the bad examples isolated exceptions to the rule?

3. If the safety committee determines that your organization has a problem, develop an organization-wide plan for solving it. The plan should result in consistent positive role modeling on the parts of executives, managers, and supervisors.

4. If the safety committee determines that most of your organization's leaders are good role models, but there are just a few who are not, develop a strategy for dealing specifically with the latter.

Endnotes

1. Isbell Construction Safety Policy. Retrieved March 3, 2008, from http://www.isbellconst.com/safety.htm.

2. Ibid.

3. Ibid.

Chapter 4

Orient Personnel to the Expected Safety-First Attitudes and Practices

Major Topics

- Importance of a comprehensive orientation
- Components of a safety-first orientation
- Traditional component of a safety-first orientation
- Safety-first component of orientation plus

The process of instilling safety-first attitudes and practices in employees should begin as part of a comprehensive orientation. Then, after the orientation has been completed, the organization should continue to communicate its safety-related expectations in a variety of ways, as explained in Chapter 2 and in the remaining chapters of this book. Most organizations provide an orientation for new employees, but the type I recommend for organizations that want to establish a safety-first corporate culture differs in several key ways from the typical orientation. This means that the safety committee will have to work closely with your organization's human resources department because the existing orientation will probably have to be revised to accommodate the recommendations contained in this chapter.

The orientation described in this chapter has two components: (1) a nonsafety component, which covers those topics typically dealt with in an orientation for new employees, but with several slight differences designed to help establish safety-first attitudes, and (2) a safety component, which is designed to help encourage safety-first practices from the outset. It is principally this second component that makes the recommended orientation different from what organizations lacking a safety-first corporate culture are accustomed to. This is the component that is most important in helping instill safety-first attitudes and establish safety-first practices. However, there are also some differences recommended in the approach your organization takes to the traditional component of the orientation. Although these differences are slight, they are important and should not be ignored.

Importance of a Comprehensive Orientation

Orientations have always been important for new employees, but in organizations trying to establish a safety-first corporate culture, they are even more important. An effective orientation makes new employees feel like part of the team, prepares them to succeed in their new job, empowers them to help the organization succeed, informs them of what is expected, and gets them started on a positive note. In addition, orientation is an important step in the process of instilling a safety-first attitude in new employees and introducing them to safety-first practices.

The following list summarizes the benefits that can accrue from the type of orientation recommended in this chapter:

- Makes new employees feel welcome and like part of the team
- Serves as a bridge that eases the transition into the organization
- Answers questions, alleviates concerns, and allays apprehensions that employees may have when starting a new job
- Lets new employees know how the workplace and your organization operate in practical terms on a daily basis
- Introduces new employees to their team members, supervisor, managers, and other key personnel
- Informs new employees of all applicable policies, procedures, practices, and other practical issues, including those relating to safety
- Allows new employees to complete—with the assistance of qualified personnel—all necessary in-processing paperwork
- Informs new employees of expectations (practical, cultural, and safety-related)
- Orients new employees to the facility (their work area, break rooms, restrooms, parking areas, supervisor's location, etc.)
- Gives the organization an opportunity to set the right tone and make a positive impression on new employees
- Begins the process of introducing new employees to the organization's safety-first corporate culture and instilling its expected practices relating to safety

To gain these benefits, many organizations will have to take a new approach to orientating new employees, an approach that goes beyond the traditional nuts-and-bolts elements covered in most orientations. This new approach should include a component that begins the process of instilling a safety-first attitude, establishing safety-first practices, and transmitting the safety-first corporate culture. I call this new, more comprehensive approach *Orientation Plus*.

Components of a Safety-First Orientation

To have the desired effect, new-employee orientation must include certain elements. Orientation Plus is a safety-first orientation because it consists of two broad components, both of which have safety-first implications: a traditional component and a safety-first component.

Although Part 2 of Orientation Plus is called the safety-first component, it is important to understand that there are also safety-related aspects of the various elements that make up Part 1, the traditional component.

Traditional Component of a Safety-First Orientation

The traditional component of Orientation Plus consists of the following elements: (1) welcome; (2) policies, procedures, rules, and employee handbook information; (3) introductions of key personnel; and (4) a facility tour. These are the nuts-and-bolts elements typically included in an orientation for new employees. However, even these traditional elements have safety-first implications, a fact that should be thoroughly understood by all involved in conducting the orientation. For this reason, your organization's safety committee will have to work closely with the human resources department and higher management in developing the type of orientation recommended herein or in revising an existing orientation to contain the necessary safety-first elements.

Welcome

Making new employees feel like part of the team is a critical aspect of Orientation Plus. Of course, this is important with any new employee, but it is even more important with new employees in organizations trying to establish a safety-first corporate culture, something everyone involved in conducting this element of Orientation Plus must understand. Employees who feel like they are part of a team of people who depend on each other to not just get the job done, but to get it done safely are more likely to develop a safety-first attitude and follow established safety practices.

Being part of a team in which individual members depend on each other is a powerful way to create and reinforce expectations. On the other hand, if new employees do not feel welcome or if they feel lost, the rest of what follows in the orientation will lose much of its impact. The way in which new employees are welcomed has implications for all aspects of an organization's corporate culture, including the safety aspects.

Introducing the Concept of Competition as Part of the Welcome   New employees need to understand how competition can affect your organization's ability to succeed as well as their corresponding job security. Consequently, as part of their welcome, new employees should be shown that the organization's success and, in turn, their own are linked directly to the concept of competition. This will provide the basis for linking your organization's competitiveness to a safe and healthy work environment later in the safety-first component of the orientation. New employees should receive a localized version of the following message as part of their welcome:

> *The better our organization performs, the better our chances of success. The performance of our organization depends in large measure on the performance of individual employees just like you, working together with your team members. Further, the more successful we are as an organization, the more likely you are to be successful in your career.*

I recommend that, during this part of the welcome, new employees be shown the organization's key performance indicators—cost, quality, service, productivity, key safety indicators—so they understand that the marketplace keeps score. I also recommend they be given a localized version of the following message:

We need your best effort, and we need you to continually improve your performance because every day people just like you in the organizations we compete against are doing their best to outperform you and, in turn, us. This is important because when our organization wins the competition, we all win in terms of job security and career growth potential. However, when we lose, we all lose in the same ways.

Later in the orientation, these concepts of competition, job security, and career growth potential will be tied to safety. The competition message in the welcome element of the orientation lays the groundwork for making that connection in Part 2 of Orientation Plus.

The importance of helping new employees understand the concept of competition and its potential impact on the organization as well as on them personally cannot be overstated. For younger employees, their new job with your organization may be the first time they have experienced a direct correlation between performance and reward. Once they understand this correlation, it will be easier for them to grasp how a safe work environment can affect performance and, in turn, rewards.

Introducing the Concept of Teamwork as Part of the Welcome The final element in the welcome portion of Orientation Plus involves introducing the concept of teamwork. Successful teamwork requires individual employees to put aside their personal agendas and work in mutually supportive ways with their teammates. Teamwork is fundamental to an organization's success. When individuals come together as a team and learn to work in mutually supportive, mutually dependent ways, a spirit of family develops. When this happens, taking care of each other becomes a high priority for teammates. This is how teamwork can reinforce safety-first attitudes and practices. A strong sense of family exists in a team that will look out for the best interests of its members—including their health and safety—and few things serve the best interests of employees more than a work environment that is safe and healthy.

I learned a valuable lesson about how teamwork can promote safety in a way that might surprise readers—during Marine Corps boot camp. Of all the organizations I am familiar with, none does a better job of conveying the mutual dependence/mutual support element of teamwork than the Marine Corps. How the Marine Corps builds teams provides an instructive example for civilian organizations trying to establish a safety-first corporate culture. The Marine Corps uses various strategies for helping new recruits grasp the concept of teamwork. Some of these are not applicable in a civilian setting, of course, but one of them is. It is called the "Crucible."

The Crucible is a grueling 54-hour test of physical endurance, mental agility, psychological stamina, and teamwork that recruits must complete on just four hours of sleep a night and limited rations. It consists of numerous challenging obstacles—called "warrior stations"—as well as a night infiltration course and 40 miles of forced marches. Teams of recruits must successfully maneuver each warrior station and complete the night infiltration course and 40 miles of forced marches before the individuals in them can earn the title "Marine."

What makes the Crucible especially valuable as a developmental tool for new recruits is the impossibility of completing any of its obstacles as an individual. No matter how fit, smart, or determined individuals are, only by working as a team can recruits overcome the various obstacles they face during this demanding week. In fact, not only must recruits work together to conquer the various obstacles that make up the Crucible, they must also figure out how best to work together.

In fact, even working well together, by itself, does not ensure success in the Crucible. Rather, recruits must study each successive obstacle and determine how best to use the abilities and resources available in their team to conquer it. For one obstacle, a given team member might possess a certain talent—strength, agility, good balance—that is especially important for success. Consequently, that team member will take the lead in helping the team conquer that particular obstacle. The next obstacle might require different capabilities, and, as a result, a different recruit might have to take the lead in helping the team conquer the obstacle.

It is easy to see how subjecting recruits to the Crucible forces them to learn how to work as a team, but what does this rigorous, 54-hour activity have to do with safety? The tie to safety comes in the form of an interesting phenomenon that occurs during the course of the Crucible, a phenomenon that illustrates a critical point about the positive effect teamwork can have on safety. As each successive obstacle in the Crucible is conquered, the teams of recruits grow closer and closer. Before long, and in most cases without even thinking about it, the recruits find themselves not just helping each other conquer the obstacles, but looking out for each other in the process, making sure they all get through the course without injury. The worst thing that can happen to a recruit is to fail to graduate with his class because of an injury sustained near the end of boot camp.

The challenges faced in the Crucible are not just difficult; some of them are dangerous. As the team members draw ever closer together with each successive obstacle conquered, making sure no member of the team is injured or hurt becomes a high priority. The attitudes of the recruits evolve without them even realizing it, and, soon, they begin to look out for the safety and health of their teammates as well as the progress of their team. This same type of mutually protective camaraderie will develop in your organization's teams when teamwork is functioning effectively.

Policies, Procedures, Rules, and Employee Handbook Information

This part of Orientation Plus covers the boilerplate information typically contained in an employee handbook and that every employee needs to know, such as work hours, pay periods, vacation, sick leave, benefits, probation period, salary/wage information, and completion of in-processing paperwork. However, there is also an important safety aspect to this element of Orientation Plus. This is where the organization's corporate safety policy should be introduced.

Since the next part of Orientation Plus involves introducing new employees to key personnel, it is possible that your organization's CEO or another executive might be available. If this is the case, you can add emphasis to the importance of the safety policy by having an executive-level manager introduce it. Having introduced the policy and, by doing so, illustrated its importance, the executive can inform new employees that the policy will be explained later in the orientation.

Introductions of Key Personnel

In this part of the orientation, new employees should receive an organizational chart. The chart should show who are the organization's key personnel and where the functional units of the new employees fit into the organization. The employees' reporting structure will then be apparent, as will where they fit in. People often feel lost when they begin a new job. They don't understand how their employer's organization fits together or where they fit into it. An organizational chart will help new employees see the big picture as well as their team's role and place in it. This, in turn, will help give meaning to each employee's new job.

In addition to the traditional aspects of this element of Orientation Plus, there is an important safety aspect. The position of the chief safety professional for your organization should appear on the organizational chart. Letting new employees know from the outset not just where they fit into the overall organization, but also where safety fits in is important. If safety does not appear somewhere on the organizational chart, the omission sends the wrong message to new employees.

There is one more benefit that can accrue from showing new employees where safety fits in on the organizational chart. It creates what I call a *teachable moment,* an opportunity to make the point that although there is a safety department or a safety professional with responsibility for leading the organization's safety program, safety is everyone's job.

Facility Tour

As with the other elements of the traditional component of Orientation Plus, there is a safety aspect to the facility tour given to new employees. People need to physically fit in as much as they need to emotionally and intellectually fit in. Consequently, it is important to include facility maps or diagrams in the orientation as well as a slow and fully explained tour of the facility.

The safety aspect of the facility tour involves the tour guide pointing out visible examples of the organization's safety features. For example, the tour guide should point out how restricted areas are marked, caution signs, the personal protective equipment (PPE) that employees are wearing, safety guards on machines and equipment, and any other visible evidence that the organization emphasizes safety.

There is more to the facility tour than just helping new employees find their way around. The tour has powerful cultural implications, especially as they relate to safety. New employees will notice how the organization's personnel go about doing their jobs. They will notice if employees wear the appropriate PPE in areas that require them. They will also notice if visitors to these areas wear appropriate PPE, such as safety glasses and hardhats. They will notice if employees heed posted warnings and follow posted directions. Consequently, it is critical that the tour guide be trained to follow established safety procedures, require new employees to do so, and be able to point out visible evidence of the organization's commitment to safety.

Safety-First Component of Orientation Plus

The safety-first component of Orientation Plus consists of the following elements: (1) the organization's safety policy and record; (2) safety-related core values from the strategic plan; (3) safety elements of job descriptions; (4) safety elements of the performance appraisal process; (5) standard PPE requirements (e.g., safety shoes, glasses, hardhats); and (6) the mentor assignment.

Your Organization's Safety Policy and Record

This might be the most important element of Orientation Plus because this is the one that says, "We are committed to a safety-first corporate culture." This is the part of the orientation that sets the tone for new employees in terms of safety-first attitudes and practices. Consequently, it should be conducted by your organization's chief safety professional or, at least, a member of the safety committee who is a safety professional.

Your Organization's Safety Policy　In this element, new employees should be provided a copy of your organization's safety policy. Then, each element in the policy should be explained with time allowed for questions and discussion. For example, the publishing giant Pearson—the parent company of the publisher of this book—has a comprehensive safety policy that contains, among other things, the following safety and health imperatives:[1]

- Provide a safe and healthy working environment
- Operate plant and equipment safely
- Train, support, inform, and encourage employees to carry out their work safely
- Provide effective arrangements to deal with work-related injuries and ill health

Your organization's safety policy was introduced during the traditional component of Orientation Plus. In this element of the orientation, it is discussed in detail so that new employees see how it applies directly to them and how it affects the way they do their jobs.

Your Organization's Safety Record　It is important to let new employees know that your organization keeps score when it comes to safety. The best way to do this is by showing them the organization's safety record covering a period of at least five years. Showing them the record makes the point that safety is so important your organization keeps score. Showing them a record covering several years makes the point that your organization monitors results and tries to improve on them over time.

Many organizations post the number of workdays or work hours lost to accidents and injuries in public locations throughout their facilities. This is an excellent practice, but what is recommended herein is even more detailed. The safety record shown new employees during this part of Orientation Plus is more comprehensive. It should contain such information as your organization's annual totals in the following areas:

- Workdays lost
- Wages lost
- Property damage

- Fire losses
- Insurance costs
- Indirect costs relating to safety and health

The purpose of providing this type of information to new employees as part of their orientation is to illustrate that (1) your organization is serious enough about safety to keep score, (2) year-to-year comparisons are made to encourage continual improvement in each reported category, and (3) poor safety costs money that otherwise could be invested in wage increases, improved benefits, incentive bonuses, new technology, better working conditions, and other issues of concern to employees.

Safety-Related Core Values from Your Organization's Strategic Plan

The point was made in Chapter 2 that an effective way to demonstrate executive-level commitment to safety is to include safety as a core value in your organization's strategic plan. In this element of Orientation Plus, new employees are shown that safety is, in fact, a core value of your organization; that it is included in the organization's strategic plan; and that its inclusion is further evidence of your organization's commitment to providing a safe and healthy workplace.

New employees should be given a copy of your organization's strategic plan. This will make the point that safety is a fully integrated organizational imperative rather than an isolated concern. All of the organization's core values should be pointed out. Then the core value relating to safety should be discussed specifically. For example, assume that a core value relating to safety in your organization's strategic plan reads as follows:

Workplace safety is an organizational imperative in our company.

The person conducting this element of the orientation—preferably your organization's chief safety officer—should explain what an "organizational imperative" is and the significance of including safety as a core value in the strategic plan. It is a good idea to allow time in this part of the orientation for questions from new employees and plenty of give-and-take discussion.

This is an excellent place in the orientation to use a fictitious case to make the point that workplace safety is an organizational imperative. When demonstrating this method for safety professionals, I encourage them to use a scenario such as the following:

Our company has a major contract with an important customer. Unfortunately, the contract is running behind schedule, and the pressure is on to meet the next delivery deadline. To speed up production, several supervisors are either advising their direct reports to remove the safeguards from their machines or looking the other way when they do so. On the basis of what you have heard about safety in our organization so far, how do you think this situation should be handled?

New employees should be allowed to suggest potential approaches for handling this situation with plenty of opportunities for discussion and questions. However, they should complete the activity understanding that deadlines are important and should be met, but in a safe and healthy manner. They are not to be met by disregarding safety procedures. The supervisors in this case should make sure that safeguards are properly fitted on all machines and that all other applicable safety practices are observed. Then, they should explore other safe tactics for meeting the deadlines, such as authorizing overtime work, pulling additional machine operators off other less pressing jobs, or calling a team meeting and asking for recommendations from their direct reports.

Your Organization's Job Descriptions

A job description can be an excellent tool for helping instill a safety-first attitude in employees provided it contains the organization's safety-related expectations. By including safety-first expectations in job descriptions, you make the expectations personal to the individual employee; they become expectations for "me," rather than for other employees.

Your organization's job descriptions should include all of the job-specific information they have traditionally contained. But, in addition, they should contain your organization's safety-first expectations. For example, a statement similar to the following would be included in all of your organization's job descriptions:

> *Individuals in this position are expected to consistently exemplify the organization's safety-first attitude and practices. An important responsibility of this position is to work safely, to expect others to work safely, and to help the organization identify and eliminate or mitigate hazardous conditions on a continual basis.*

Repetition is important when trying to instill a safety-first attitude in new employees. Remember, the concept might run counter to their natural inclinations. This is why it is important for new employees to hear about your organization's safety-related expectations in every element of the safety-first component of Orientation Plus and to see them consistently modeled by executives, managers, and supervisors.

Your Organization's Performance Appraisal Process

There is a management axiom that says: "If you want performance to improve—you have to measure it." Any time you have expectations, it is important to determine the extent to which employees are meeting them. In other words, if you have expectations relating to safety (or anything else), measure the actual performance of employees against the expectations. This probably sounds like a statement of the obvious to experienced safety professionals. However, it needs to be said.

My experience as a consultant has repeatedly shown that a gap often exists between what is expected of employees and what is measured. In practical terms this means that the core values in the strategic plan, safety policy, job description, and other management tools expect one thing, but the performance appraisal instrument measures another.

Expectations communicated to employees through any element of Orientation Plus, especially the safety-first expectations, should become part of the criteria against which employee performance is measured during the performance appraisal process. This might mean that your organization will have to revise its existing performance appraisal instrument(s) and usually does. This, in turn, means that the safety committee will have to work closely with your organization's human resources department.

Any revisions required should be completed before distributing the performance appraisal instrument during the orientation so that the instrument becomes another useful management tool for conveying safety-first expectations. There are two approaches for including safety-related criteria in your performance appraisal forms: one broad, the other specific.

Broad Approach   With this approach, one broadly stated criterion is added to your organization's performance appraisal instrument that covers all of your safety-first expectations. An example of such a criterion follows:

- This employee meets all safety-related expectations.

> CT = Completely True
> ST = Somewhat True
> SF = Somewhat False
> CF = Completely False

Of course, the rating method used in this example is just that—an example. The rating method currently used in your performance appraisal instrument will probably work just as well provided the wording of the criterion itself is appropriately adapted.

With this approach, if an employee receives any rating except "Completely True" for the safety criterion, shortcomings relating to specific safety-first expectations should be recorded in the *comments section* of the performance appraisal instrument and explained by the supervisor during the appraisal follow-up conference with the employee in question.

Specific Approach   With this approach, the specific expectations listed in employee job descriptions are included as criteria in the performance appraisal instrument. For example, the following safety-related criteria might be included in your organization's performance appraisal form:

> CT = Completely True
>
> ST = Somewhat True
>
> SF = Somewhat False
>
> CF = Completely False

- This employee maintains a safety-first attitude
- Exemplifies safe works practices

- Encourages other employees to work safely
- Helps identify and eliminate or mitigate hazardous conditions in the workplace
- Contributes to the maintenance of a safety-first corporate culture in our organization.

Showing new employees, as part of their orientation, that their performance will be evaluated and that the evaluation will include safety-related expectations sends a powerful message about the importance of these expectations. Remember, most employees have been conditioned by school and life to understand that what is evaluated is important. Consequently, they need to see from the outset that, in your organization, their job-performance measures include safety. The performance appraisal instrument, if properly developed, will send the right message.

Your Organization's PPE Requirements

If your organization has PPE requirements, they need to be provided to new employees during this element of Orientation Plus. Of course, with most organizations, some employees are required to use certain types of PPE and others are not. Individual supervisors can explain the specific PPE requirements to their new direct reports during the team-level orientations they provide. At this point, what is important is organization-wide PPE requirements.

Are all employees required to wear safety glasses? Safety shoes? Hardhats? If so, this should be explained in this segment of the orientation. If PPE must be worn only in certain areas of the facility, this should be explained as well as which areas and where the required PPE is located. In this element of Orientation Plus, employees should learn whether they are required to wear PPE, when, under what conditions, and how to locate it. Its proper use can be explained during team-level orientations conducted by supervisors with the assistance of safety professionals.

Your Organization's Mentor Assignments

New employees must absorb a lot of information during any orientation, but especially during one as comprehensive as Orientation Plus. Consequently, chances are that employees will feel overwhelmed by all they have seen and heard. This is where your organization's mentors come into play.

To ensure that new employees get off to a good start, it is helpful to assign them a mentor at the conclusion of their orientation. There is nothing new about assigning mentors to new personnel. Using experienced personnel to help new employees get a good start is a time-tested and effective management strategy. In fact, Chapter 5 covers in detail the use of mentors for helping new employees develop safety-first attitudes and work practices.

The approach recommended herein for using mentors as part of Orientation Plus may be new to your organization. In this approach, a mentor is assigned to each new employee at the conclusion of the employee's orientation. The mentor should be the personification of your organization's cultural and safety-first expectations. He or she need NOT be a technical, job-related mentor. In fact, the mentor assigned at this point may even be in a different job and from a different department. On the other hand, the mentor assigned at the conclusion of the orientation can be from the same department, work in the same job, and serve as both a job-related and cultural/safety-first mentor depending on the capabilities of the individual in question.

Regardless of the approach taken, the postorientation mentor's task is to reinforce the cultural and safety-first expectations presented during orientation and to help new employees begin to accept and internalize these expectations. The postorientation mentor's job is to make sure that what was presented during Orientation Plus is not forgotten, ignored, or discarded. The mentor will typically work with new employees during that all-important first month when they will be forming work habits, which should be proper from the outset.

New employees should meet with their mentor daily before work, at lunch, or at the end of the day to discuss their impressions, questions, problems, and issues relating to how things are done in your organization. In addition, the mentor should be available at all times by telephone, e-mail, and/or other means of electronic communication. The mentor should reinforce the right way of doing things and point out any areas in which the new employee may be falling short. The mentor serves as a role model, an ear to listen, and even a shoulder to cry on, but his or her ultimate goal is to ensure that employees successfully bridge the gap between where they are when they show up for their new job and where your organization needs them to be in terms of safety-first expectations as well as all of your organization's other cultural imperatives.

SAFETY-FIRST CORPORATE PROFILE
ALCOA

ALCOA is one of the world's largest aluminum producers. Founded in 1888, ALCOA has its principal headquarters in New York City and its operational headquarters in Pittsburgh. ALCOA's product line includes fastenings, castings, building products, aluminum foil, automobile parts, rolled aluminum, and milled aluminum. ALCOA is a $30 billion plus company that employs approximately 129,000 people in 44 countries. ALCOA competes on a global scale and operates aluminum smelters in 25 different locations.

Aluminum smelting is a potentially hazardous operation. Consequently, ALCOA must take employee safety and health seriously, and it does. Safety, health, and environmental management are part of ALCOA's broader Global Ethics & Compliance system. The system, as it relates to safety, health, and environmental concerns, consists of specific performance metrics that are monitored in all ALCOA facilities worldwide to ensure that the company's operations and processes do no harm to people, communities, or the environment. As a result of its commitment to safety, health, and environmental management, ALCOA is considered one of the safest companies in the world.

In addition to establishing high internal expectations relating to safety, health, and environmental management, ALCOA requires its suppliers to comply with strict standards in all of these areas. ALCOA requires that its suppliers adhere to the same safety, health, and environmental value system the company has committed to. There is a lesson in this for other organizations. A safe company that does business with unsafe partners is not totally committed to safety.

Source: http://ethisphere.com/2007-worlds-most-ethical-companies/

There are several strategies that can help organizations establish a system for using mentors to help instill a safety-first attitude. Implementing these strategies will require the safety committee to work closely with your organization's human resource department and to secure approval for some strategies from higher management. The first strategy is to provide those chosen to serve as mentors a stipend that is over and above their normal salary. Being a mentor is extra duty. Consequently, those who serve as mentors should be adequately compensated for their time, effort, and expertise.

The second strategy is to provide those chosen to serve as mentors the training necessary to ensure that they (1) thoroughly understand the organization's cultural and safety-first expectations, (2) know how to exemplify those expectations in their attitudes and practices, and (3) are able to communicate effectively one-on-one with the new employees they mentor.

The type of orientation recommended in this chapter for new employees—Orientation Plus—is different and more comprehensive than the typical orientation. This is as it must be if your organization is going to get off to a good start instilling a safety-first attitude in new employees, an attitude that will lead to safe work practices day in and day out on the job. A comprehensive, effective orientation provides a firm foundation for all of the remaining steps in the model presented in this book.

COLLEAGUE-TO-COLLEAGUE DISCUSSION CASES

CASE 1: The Insufficient Orientation

Jake Hess is the safety manager for Tech-Plus Corporation (TPC), an engineering and manu-facturing company of almost 3,000 employees. TPC has an acceptable safety record—not good, but acceptable. Hess would like the company to do better. He would like to lead the way in transforming TPC into an exemplary model of a safety-first company, and he is making progress. However, gaining buy-in from higher management has taken Hess almost two years because the company has experienced no major accidents or disasters that might focus the attention of higher management on safety. The attitude of TPC's executive team toward safety has been, "No news is good news."

To Hess, the organization's orientation for new employees is the perfect example of what ails TPC in the area of safety management. The company's orientation deals with all employee concerns except safety. There is not even one safety-related element included in the orientation. Consequently, Hess has decided to begin his efforts to transform TPC into a safety-first company with the company's orientation. The safety committee agrees with Hess, but nobody seems to know how to get started.

Discussion Questions

1. Does your organization's orientation for new employees contain a safety compo-nent? If so, how does it compare to the safety-first component of Orientation Plus presented in this chapter?

2. If Jake Hess asked you and your colleagues how to go about getting his company's orientation revised to reflect a safety-first commitment, what would your group advise?

Case 2: The Reluctant HR Director

Patricia Martan is the director of safety and health for Reynolds Corporation. Her colleague Jack French—the company's director of human resources—has become a stumbling block in Martan's efforts to establish a safety-first corporate culture at Reynolds Corporation. French has been reluctant to revise the company's orientation for new employees, but he has gone along—albeit reluctantly—with several of Martan's suggestions. However, she is having trouble making him understand why the concept of competition should be explained as part of the orientation.

Martan used competition and compliance as the carrot and stick of her rationale when convincing higher management to commit to a safety-first corporate culture. French understands that higher management is committed to employee safety and health. In fact, he claims to share this commitment. However, when it comes to making changes in his area of responsibility, French always seems to find ways to slow down the progress Martan is making. His questioning of why the concept of competition should be explained as part of the orientation is just his latest tactic for slowing the transformational process.

Discussion Questions

1. Have you or your colleagues ever had to deal with another manager whose cooperation you needed but had difficulty securing? If so, discuss the situation you faced and how you handled it.

2. Put yourself in Martan's place. How would you explain the importance of including the concept of competition in new-employee orientations?

Key Terms and Concepts

Before leaving this chapter, make sure you understand the following key terms and concepts and can accurately explain them to people who are not safety professionals.

Comprehensive orientation

Orientation Plus

Traditional component of a safety-first orientation

Safety component of Orientation Plus

Review Questions

Before leaving this chapter, make sure you can accurately and comprehensively, but succinctly, answer the following review questions.

1. Explain the importance of a comprehensive orientation for new employees.

2. List the traditional components of a safety-first orientation.

3. Select those elements from the list in Question Number 2 that have safety ramifications, and explain those ramifications.

4. Explain how each of the following elements of the safety component of an organization's orientation can be used to help establish a safety-first corporate culture:

   a. Safety policy and record

   b. Safety-related core values from the strategic plan

   c. Job descriptions

    d. Performance appraisal process

    e. PPE requirements

    f. Mentor

Application Project

Revising your organization's orientation is just one of many situations that will require you and the safety committee to work closely with other departments and personnel in order to implement the transformational model presented in this book. Before approaching your organization's human resources department about including safety in the orientation for new employees, the safety committee should develop what I call a "strawman." A strawman is a draft of a plan showing what you would like the orientation to cover. It is developed as an annotated outline containing all of the elements explained in this chapter. Working with your safety committee, develop a strawman for a version of Orientation Plus that could be implemented in your organization.

Endnote

1. Pearson Health and Safety. Retrieved April 14, 2008, from http://www.pearson.com/index.cfm?pageid=148.

Chapter 5

Provide Mentors Who Exemplify Your Organizaton's Safety-First Expectations

Major Topics

- Cultural/safety-first purpose of the mentor
- Mentoring strategies for having a positive influence
- Best practices for giving feedback to protégés
- Encouraging protégés to internalize the safety-first corporate culture

The importance of providing a comprehensive orientation for new employees that includes a strong safety-first component was stressed in Chapter 4. The concept is called "Orientation Plus." This type of orientation is an important step in the development of a safety-first attitude and work practices. However, even when done well, Orientation Plus, like all orientation programs, has an Achilles' heel; its weakness is that it requires new employees to absorb a lot of information in a short period of time. An effective way to overcome this weakness is to assign new employees a mentor at the completion of their orientation.

The mentor is an experienced individual who can help new employees absorb what was presented during orientation and put it to good use in their new jobs. The most important responsibility of the mentor is facilitating the new employees' internalization of your organization's corporate culture and safety-first expectations. Consequently, people assigned as mentors should be exemplary role models of your organization's cultural expectations.

Kellogg's is just one of many corporations that assign mentors to help new employees absorb, adopt, and apply their cultural values. For example, one of the corporate values at Kellogg's is a safety-first attitude that results in safe work practices. Kellogg's wants its employees to work hard, work smart, and work safe. One of the ways the corporation's safety-related values are put into action in practical, everyday terms is the company's mentoring program.

Cultural/Safety-First Purpose of the Mentor

A mentor is an enabler and a facilitator, an experienced individual who helps others succeed. In the current context, the mentor is someone who helps new employees gain a sufficient foothold in the organization to become fully contributing, fully accepted members of the team, members who understand, accept, and internalize the organization's cultural and safety-first expectations. The mentoring process might involve answering very practical questions related to rules, processes, benefits, and so on, but the emphasis of the mentor in the current context is on facilitating the protégé's acceptance and application of the organization's corporate culture and safety-first expectations.

If your organization already has a mentoring program, it will probably have to be revised to include the safety-first aspects. If your organization has no mentoring program, it will need to start one. In either case, the safety committee must be prepared to take the lead in (1) revising or developing as appropriate a mentoring program that includes safety-first responsibilities, (2) securing the approval of higher management for the new or revised mentoring program, (3) developing and presenting a comprehensive training program for prospective mentors, and (4) working with the organization's supervisors and its human resources department to effectively implement both the training and mentoring program.

The ideal mentor for helping facilitate the internalization of safety-first values is a member of the new employee's team who has internalized the organization's culture and is an exemplary role model of its safety-first expectations. However, mentors do not have to come from the new employee's team. They can come from other teams and even other departments. Before assigning personnel to serve as mentors, make sure they understand the purpose of the mentoring process and the specific duties they will be responsible for carrying out. The best way to do this is to develop a mentor training program, as mentioned above. This chapter contains all of the information needed to develop such a program.

Purpose and Duties of the Mentor

As was stated earlier, mentors are responsible for helping new and less experienced employees make a good start toward becoming fully contributing, fully accepted members of the team who understand, accept, internalize, and apply the organization's corporate culture and safety-first expectations. Specific duties of the mentor include the following:

Making New Employees Feel Welcome   Most people have an inherent need to fit in, and this need is felt acutely by new employees. Until they feel welcome and part of the team, new employees will be unable to focus sufficiently on their performance, including safety-first work practices. From the perspective of safety, a distracted employee is a dangerous employee. It is difficult enough to focus sufficiently on learning to do a new job and do it safely under even the best of circumstances. When distracted by a feeling of not fitting in, it can be even more difficult.

Think about this phenomenon of feeling like the odd person out. Have you ever attended a social function where everyone except you knew everyone else? If so, you know how uncomfortable it can be. This is how new employees feel until they are welcomed into the team and made to feel a part of it. An employee who feels like an outsider will not have the focus needed to learn and apply safe work practices and, as a consequence, can be a danger to himself and others.

The good news is that just assigning new employees a mentor is a major step toward making them feel welcome. Mentors can enhance the welcome by using such tactics as (1) asking their protégés to join them for lunch (protégé is the term given to those who are being mentored); (2) introducing protégés to other employees—even those outside of their department; and (3) calling, e-mailing, or stopping by to see protégés several times a day to ask "how are you doing?" More than anything, a mentor can be a friendly, familiar face the new employees can approach when questions, doubts, concerns, or problems arise.

The flipside of making new employees feel welcome is making them understand that they are, in fact, part of a team, rather than an island unto themselves in a sea of other employees—an important point for new employees to understand. Mentors can use the inherent need to fit in as a way to begin promoting a safety-first attitude and safe work practices.

Helping New Employees Understand How Safety Fits Into the Organization's Strategic Direction   As was shown in Chapter 4, a comprehensive orientation in the mold of Orientation Plus will include an explanation of the safety-related corporate values in the organization's strategic plan. The strategic plan gives new employees a sense of the organization's big picture. The mentor can help new employees understand how safety fits into the organization's strategic direction.

In addition to showing them where safety fits into the organization's big picture, mentors can help their protégés understand what the organization's safety-related core values mean to them in practical terms. For example, the mentor might show a protégé an important process and, while explaining why the process is important, also point out the various safety precautions those who operate the process are taking. This is an excellent way to show that safety and productivity are just two different sides of the same coin in your organization.

Giving New Employees a Visual Example of the Safety-First Corporate Culture   Providing a visual example amounts to being a good role model for protégés and is perhaps the most important responsibility of the mentor. Mentors should be "recruiting posters" for the organization's safety-first corporate culture. If mentors fail in this responsibility, they are sending a powerful message that says: "All of this talk about safety-first is just that—talk." The mentor's attitude and work practices must exemplify the organization's most important cultural expectations.

Once, many years ago when beginning a new job, I was assigned a mentor. Why this individual was asked to serve as a mentor is still a mystery to me because he was a terrible example for new employees. As soon as my orientation was completed, this mentor told me to "forget all of that nonsense. I will show you how things really work around here." He then spent all of our time together demonstrating how to cut corners, ignore regulations, and goof off on the job without getting caught. His whole approach to work was to do as little as possible while appearing to be busy. By this time in my career I had enough experience to know that I should ignore the advice of this so-called mentor, but I hate to think what would have happened had this been my first job. Mentors should be chosen with care—they serve a critical purpose.

Helping New Employees Overcome the "Rookie Jitters"   New employees go through an inevitable period of what I call the "rookie jitters." Until they gain a sufficient foothold to feel comfortable in their new surroundings, newly hired employees are likely to be nervous, anxious, and uncertain. From the perspective of safety, this is the worst possible frame of mind for an employee. Helping protégés work through the rookie jitters is an important responsibility of the mentor.

Until new employees have worked through their concerns and issues, they will be unable to focus on learning to do their jobs properly—which means effectively, efficiently, and safely. The following is an important message that can be sent by the mentor when helping a protégé work though a concern or an issue: "Here is how we handle that situation in this organization." For example, when the new employee first feels the stress of an impending work deadline, he or she might respond by ignoring important safety precautions. This situation gives the mentor an excellent opportunity to demonstrate what *safety-first* means on a practical level by suggesting other ways to meet the deadline while still observing all applicable safety precautions.

If the company in question is ABC, Inc., there is an ABC way of dealing with the kind of problem this new employee is facing, a way that is consistent with the company's cultural and safety-first imperatives. The mentor's job is to help protégés understand the ABC approach to handling the problem he or she faces. The new employee needs to understand that although deadlines are important and must be met, there are ways to catch up when you fall behind other than ignoring applicable safety precautions. This is the message the mentor should give the protégé in such a situation. It is an important message for new employees who are still unsure of how your organization functions on a practical level. It tells them that when they confront a problem or face a challenge in their work, the way to deal with it is the ABC way, and the ABC way is the safe way.

Reinforce the Significance of the Safety-First Component of the Orientation   When having lunch or just chatting when on break, the mentor should reinforce the cultural and safety-first information that was introduced during the new employee's orientation. For example, mentors can explain further why certain customs, traditions, rules, and regulations are important and the significance they have in the organization. Mentors might expound further on the organization's successes in the area of workplace safety and give actual examples that bring these successes to life for protégés.

Hearing about safety-first successes and their significance to the organization's competiveness in one-on-one conversations with a mentor will often make even more of an impression than the one made during the orientation. People often respond better to guidance that is provided in this type of informal, less threatening setting than to the approach that is necessarily used in a group orientation. One-on-one, informal discussions are one of the best ways for mentors to help new employees develop the desired attitudes and practices relating to safety.

Mentoring Strategies for Having a Positive Influence

Mentors have an unparalleled opportunity to shape the attitudes and work practices of protégés in a positive way, but the opportunity will be realized only if mentors have the necessary influence with their protégés. The following strategies will help mentors establish an influential relationship with their protégés.

Be an Active, Patient Listener

Good listening skills are essential when working with any employee, but they are especially important when working with new employees. Remember, new employees need the time and attention of an experienced person, in this case the mentor, even more than other employees. Probably the most helpful thing a mentor can do for a protégé is to listen.

There are several reasons why it is important for mentors to listen to their protégés: (1) the protégé might be frustrated and in need of an opportunity to vent, (2) the protégé's problem might be evidence of an even larger problem within the organization, (3) the protégé might be having a problem with another employee who is not properly modeling the organization's safety-first corporate culture, (4) the protégé might be confused about some work-related issue, and (5) the protégé's problem might reveal the need for some type of improvement.

Clearly, mentors must be good listeners. What follows is a list of tips that will help improve any person's listening skills. These tips are especially important for individuals who serve as mentors.

- Maintain a positive, nonjudgmental attitude, tone of voice, and facial expression.
- Maintain eye contact with the protégé throughout conversations.
- Eliminate distractions and give the protégé your undivided attention.
- Let protégés explain their problems and concerns without interruption.
- Observe nonverbal cues to determine if what the protégé "says" nonverbally matches what he or she says verbally.
- Keep an even temper no matter what is said (mentors who lose their temper with protégés will also lose their credibility and influence).
- Make mental rather than written notes as you listen (taking written notes while a protégé is talking can be distracting to him or her).
- Paraphrase and repeat back what you think you heard the protégé say.
- Ask questions to clarify anything you do not understand and to gain more information if necessary.

Maintaining a positive, nonjudgmental attitude will encourage protégés to share what is on their minds and to be open, frank, and thorough. Negative nonverbal cues such as frowns or looks that say "I don't want to hear it" or "I don't have time for this" will discourage protégés. Maintaining eye contact with protégés lets them know you are listening, and it will help promote better communication in both directions.

It is also a good idea to eliminate distractions that might pull your attention away from the protégé. The most common distractions in these situations are typically pressing items of work on your desk and cell phones. Rather than eliminate desktop distractions by going through the tedious and time-consuming task of cleaning off your desk, try putting two chairs in front of your desk—one for the protégé and one for you. This will eliminate your worst distractions by putting them behind you. As for your cell phone, just turn it off temporarily.

New employees can be frustratingly uninformed. Even so, it's best to let them talk until they have fully expressed their thoughts, without interruption. Interruptions undermine the protégé's momentum and should, therefore, be avoided. You can use nonverbal cues to improve your listening. Observe protégés closely as they talk. Do their nonverbal messages match their verbal message? If not, remember to ask questions to clear up the inconsistencies once protégés have said what they have to say.

Perhaps the hardest thing to do when listening to a protégé is to maintain your temper when something that is said makes you angry, which is going to happen from time to time. Because of their newness protégés can sometimes be frustrating to mentors, and frustration can lead to anger. Mentors must remember above all else that an angry response to a protégé's comments will not just undermine communication, it might damage the mentor's image and, in turn, credibility and influence with new employees.

It is only natural when listening to someone to want to take notes. However, doing so can have unintended consequences such as distracting the person who is talking. New employees talking to a mentor might begin to wonder what the mentor is writing down rather than concentrating on what they need to say. Consequently, it is better to just listen. Mentors can make contemporaneous notes once the conference is over, if necessary.

Paraphrasing and repeating back to protégés can be beneficial in two ways. First, when mentors can accurately and succinctly summarize what their protégés have said, they convey that they have been listening. Second, if the mentor has misunderstood what was said, paraphrasing gives protégés an opportunity to correct the misperception. Finally, once protégés have said what they need to say, clarifying questions should be asked to fill in any gaps in the mentor's understanding and to provide any missing information. In this way mentors can avoid acting on the wrong perception or misinformation.

Work to Establish Trust

People will not follow, believe in, or be positively influenced by those they do not trust. Trust is the foundation of influence, and mentors need to be influential with their protégés. Hence, they need to be trusted by them. Consequently, any time spent by mentors in building trust with protégés is time well spent. One of the best ways for mentors to build trust is to share stories about the insecurities and concerns they felt as a new employee and any they might still feel. The unstated message behind such stories is "if I am willing to trust you with my insecurities, you can trust me with yours."

This is important because when mentors stress the point that the safe way is the right way, they need their protégés to accept this as fact. A protégé is just like you. If you trust someone, you are more likely to believe and accept what he or she tells you. If you do not trust a person, it is easy to disregard what he or she says.

Another way to build trust with protégés is to show a sincere interest in them and their success. People can be very perceptive. They can sense when mentors are just reluctantly doing their duty or just going through the motions to complete an unwanted assignment. On the other hand, they can also sense when mentors are sincerely interested in helping them. When protégés sense a sincere desire to help and a caring attitude, they are more likely to trust their mentor. This trust, in turn, will enable the mentor to influence protégés in a positive way.

Be Empathetic

Mentors should try to remember how they felt as rookies. Were they anxious? Frightened? Shy? Overwhelmed? Confused? As new employees, mentors probably felt the same anxieties their protégés are feeling now. Being able to look protégés in the eye and say, "I felt that way too when I was a new employee," lets them know that (1) their concerns are valid, (2) they are not the only person to have these concerns, and (3) the anxieties and concerns will eventually pass, just as they did for their mentor.

Be a Consistent, Positive Role Model

People who say "Do as I say, not as I do" are rarely able to influence others in a positive way because they lack the necessary credibility. The best way to gain credibility with new employees is to be a consistent, positive role model of the organization's cultural expectations. As a mentor, let them see you living out the values professed by your organization.

SAFETY-FIRST FACT

Workers' Compensation Claims Are Higher in the Food Service Industry

A study entitled "Safety & Health Assessment & Research for Prevention" (SHARP) found that compared with other industries, food processing has a high rate of workers' compensation claims. However, the study also revealed that food-processing companies that (1) were organizationally healthy had lower job-related injury and illness rates and (2) had systems in place to keep workers safe and healthy had lower job-related injury and illness rates.

Source: Retrieved from http://wa.gov/Safety/Research/HealthyWorkplaces/Food/default.asp on March 23, 2008.

I saw an excellent example of putting this concept into action when I joined the Marine Corps many years ago. Knowing how important it is that its primary mentors for new recruits exemplify its culture, the Marine Corps goes to great lengths to make sure that its drill instructors, who are the primary mentors for new recruits, are the walking personification of everything that is important in the Corps' culture and everyday work practices. Drill instructors are perfect role models of everything the new recruit is expected to do. The way the recruit is supposed to stand, walk, march, dress, eat, shoot, talk, run, climb, and think is the way the drill instructors do it, except that the drill instructors do it with apparent effortlessness, do it better, and do it consistently. This fact makes a powerful impression on Marine recruits and is one of the Corps' most effective strategies for instilling its corporate culture and attitudes toward work.

It is just as important for mentors in your organization to be exemplary role models. They need to be the absolute personification of your organization's cultural and safety-first expectations. The value of the impression made on a new employee by a mentor who is an outstanding role model cannot be overstated. All employees, but especially new ones, respond better to action than words.

Show Sincere Interest in Protégés

People can sense intuitively when someone is sincerely interested in them and their problems, concerns, issues, and success. This is especially the case with new employees—people for whom a heightened awareness is a given. Mentors who are perceived as just putting in their time with protégés or just checking off an unwanted assignment will not be able to influence their protégés in a meaningful way.

However, those who show that they are sincerely interested in giving their protégés a good start will be able to have a positive influence even if they fall short in some of the other areas presented in this chapter. This is because when people know that someone truly cares about them, they are usually willing to overlook shortcomings that would otherwise be problematic. The best mentors truly take pride in the success of their protégés, and the protégés know it. Mentors who take pride in the success of their protégés send an important message to new employees, a message that says "this organization cares about you, or it would not assign such a caring mentor to help you."

Best Practices for Giving Feedback to Protégés

Giving effective feedback is one of the most important things mentors can do for their protégés. It lets new employees know how they are doing and gives them opportunities to learn the right way—meaning the safe way—from the outset. Feedback applies to what new employees are doing right as well as what they need to do better or just differently. Feedback can be an excellent tool for helping new employees understand, accept, and internalize the organization's safety-first corporate culture, but only if it is given effectively.

It is important to remember that new employees are likely to be insecure—the rookie jitters explained earlier. Their insecurity might cause them to be averse to constructive criticism. Because of the insecurity that is inherent in being new, feedback that is poorly given can do protégés more harm than good. What follows is a list of strategies mentors can use to make the feedback they give protégés more effective.

Be Encouraging—Begin with Something Positive

New employees who are insecure often take constructive criticism as just plain criticism. Consequently, how constructive criticism is given can be important. If given ineffectively, the constructive aspect of it can get lost in the protégés' hurt feelings and heightened insecurity.

One way mentors can overcome this potential problem is to introduce their constructive criticism with a few positive statements about something their protégés are doing well. For example, consider the following two approaches that might be used to tell a new employee he needs to improve in the area of punctuality:

First Approach: "John, you were late getting to work twice this week. You need to get to work not just on time, but about 15 minutes early. When I say work begins at 8:00 A.M., it's a requirement, not a suggestion."

Second Approach: "John, I was proud of the way you spoke out in the team meeting last week. It showed that you have some good ideas and are willing to get involved in making improvements. Now I want you to concentrate on doing an equally good job with punctuality. You were late getting to work on Tuesday and Thursday of this week. Punctuality is an important part of the corporate culture in our company. You really need to get to work not just on time, but about 15 minutes early."

In the first example, the mentor's feedback is intended to be constructive. The mentor is just trying to let the protégé know that punctuality is an important expectation in the organization. However, John—being human—might respond by being defensive or getting his feelings hurt because of the tone of the message. Without saying it, he might think, "OK. That's what I am doing wrong. Am I doing anything right?"

In the second example the mentor compliments John on something he has done well before letting him know where he needs to improve. With this approach, the protégé gets some positive reinforcement before hearing the constructive criticism. A seemingly small detail like this can be especially important when dealing with new and insecure employees.

One of the better definitions of tact is that it means hammering in the nail without breaking the board. Another definition of tact is making your point without making an enemy. When giving constructive criticism, tact is important, especially with new employees who are just beginning to get acquainted with the way things are done in the organization. Tact will help mentors make their point without making an enemy, hurting feelings, and beating down a new employee, who probably still feels overwhelmed with the newness and unfamiliarity of the situation.

Think About What You Are Going to Say Before Saying It

In human interaction, how something is said can be just as important as what is said. This is especially true when what is going to be said amounts to "You need to do better." As was stated above, tact can be viewed as hammering in the nail without breaking the board or as making your point without making an enemy. Both definitions suggest the same advice for mentors: Pay attention to what you say to protégés and how you say it. Before giving feedback, think about how it might be perceived by the protégé.

One way to find a tactful approach for giving feedback is for mentors to put themselves in the protégé's place. When they were new employees, how would they have wanted their mentor to communicate the message "you need to do better"? One of the most common mistakes made by mentors is rushing to give feedback so they can get on with other pressing obligations. Feedback given in a rush or without careful thought can do more harm than good. Consider the following approaches for giving the same feedback:

Approach A: "Jane, that dress is not appropriate. You read the dress code in the employee handbook. This is work, not a cocktail party."

Approach B: "Jane, that is a lovely outfit. In fact, it will be perfect for the company's Christmas party. But for daily work attire, you might want to dress a little more conservatively. I probably should have been clearer concerning the company's dress code."

In both examples, Jane was given feedback explaining that her outfit was inappropriate. However, the first example—although it makes the point—is tactless. Unless Jane is willing to give her mentor the benefit of the doubt—which new and insecure employees may not be willing to do—she will probably be hurt by the mentor's comment. The second example makes the point that needs to be made, but it does so in a tactful way. In fact, the mentor really makes the feedback easy for Jane by accepting responsibility for not being clearer about the dress code.

Give Feedback Frequently

Feedback should never be saved up for later. Rather, it should be given frequently on a just-in-time basis. There is a principle of learning that applies regardless of whether the learner in question is a child or an adult. It says: "Learners need immediate and continual feedback." This principle applies to new employees who are being mentored just as much as it applies to participants in a classroom or seminar, because protégés are learners.

People who are in a learning mode need to know how they are doing, and the sooner and more frequently they are told the better. This is especially the case with new employees. Think about how you felt in the past when taking a test in school. The first thing you wanted to know after completing the test was how you did—your score. As a student you probably did not appreciate an instructor who made you wait for days or weeks before returning the graded test. Protégés are like that—they want and need to know how they are doing.

Regular feedback need not take the form of a systematic evaluation. Some of the best feedback a mentor can give a protégé is a pat on the back that says "good job." But remember that new employees are learning. Consequently, it is important to give accurate and honest feedback. Mentors should never make the mistake of reinforcing a negative attitude or behavior in a protégé by giving artificial praise. Give praise when it is appropriate. When it is not, provide corrective feedback and encouragement.

There is another important reason for giving frequent and continual feedback: People have a tendency to habitually do things the way they initially learn to do them. In other words, the way people do certain things can become a habit. This is good if they learn the right way from the start, but it can be a real problem if they learn the wrong way. Consequently, it is important to correct new employees who are doing things wrong before the wrong way becomes habitual. This is especially true as it applies to safety-related work habits.

Find Protégés Doing Things Right, and Reinforce Those Behaviors

We tend to think of giving feedback as correcting someone, and, of course, this is often the case. But feedback should not be limited to the corrective kind. It can also take the form of positive reinforcement of appropriate behavior. Mentors should be vigilant in finding their protégés doing things right and then letting them know. Mentors who observe protégés behaving in accordance with the organization's safety-first expectations should reinforce their positive behavior right away and in specific terms. In fact, receiving this type of positive feedback will make receiving the inevitable corrective feedback more palatable for protégés.

Make Sure Protégés Understand What You Are Telling Them and Why

Human communication is an imperfect process at best. Just because you tell protégés something and even though you know they heard you, do not assume they understand what you mean. There are just too many variables that can distort what you think is a clear and understandable message. What follows is a list of common inhibitors of communication:

- Differences in meaning in which the same word can have different meanings for different people
- Insufficient trust between mentor and protégé
- Information overload
- Interference
- Condescending tone from the mentor
- Poor listening skills on the part of the protégé or the mentor
- Premature judgments and inaccurate assumptions by the protégé or mentor

People of different backgrounds, cultures, and ages can affix different meanings to the same words and nonverbal gestures. For example, when hearing a suggestion or recommendation, a young employee might say, "I'm down." By this she means that she agrees or is onboard with the recommendation. However, to a 40-something colleague, "I'm down" is more likely to mean "I'm sad or feeling blue." Differences in meaning such as this can arise between mentors and protégés, distorting the message and, in turn, inhibiting effective communication.

Trust, or the lack of it, can inhibit communication. If protégés do not trust their mentor, all messages from the mentor will have to pass through a filter of skepticism. When this happens, instead of listening for the message, protégés will tend to listen between the lines for hidden messages or for what the mentor really means but is not saying. While distracted in this way, protégés might miss or just misinterpret the intended message.

Information overload is always a potential issue when dealing with new employees. They have to take in so much so fast that it can be overwhelming. Remember, this is one of the reasons for assigning mentors in the first place. Interference can also inhibit communication. Any kind of outside interference that occurs when feedback is being given can distort the message. The simplest form of interference is noise. This is why feedback should be given, to the extent possible, in an environment free of noise and other distractions.

Mentors who are condescending in their attitudes run the risk of having their feedback misunderstood or rejected simply because people do not like to be talked down to. Talking down to protégés can cause them to tune out the message being sent. Further, it can drive a wedge of resentment between protégés and their mentors.

Listening is always an issue in any form of verbal communication. If protégés are poor listeners—and they are likely to be since many people are—the mentor's feedback may be incompletely received or inaccurately perceived. In fact, because new employees can feel so overwhelmed by all of the information being presented to them in so short a time, they might go into overload mode and simply tune out what is being said by mentors. Consequently, it is a good idea for mentors to occasionally stop and ask their protégés to paraphrase what has been said and state their perception of it in their own words.

If protégés fall into the habit of making premature judgments or inaccurate assumptions about the feedback mentors are giving them, the message received might be distorted. This can happen, for example, when protégés think they are about to hear again something they have already heard. If they make a premature judgment or an inaccurate assumption about the impending feedback, protégés might not listen and, as a result, will miss the real message. Because of the predictable inhibitors of effective communication explained here, mentors must understand that using corrective feedback to help new employees improve will take time, persistence, and patience.

Be Specific and Give Examples

When giving feedback to protégés, regardless of whether the feedback is complimentary or corrective, mentors should be specific and give examples. For example, when providing corrective feedback to a new employee who seems to have a bad attitude about safety regulations, be specific concerning how that attitude manifests itself and why it is wrong. It is important for new employees to understand that mentors know what they are talking about because some of them—new employees—will have a knee-jerk defensive reaction to corrective feedback. Such people will sometimes try to deny any claim made by a mentor that cannot be clearly backed up with specific examples.

Because of the potential for defensiveness on the part of protégés, mentors should never leave room for denials. In addition, it is important that protégés fully understand what they are being told. Feedback that is specific and that is illustrated by examples of actual behavior will promote understanding and, in turn, facilitate the developmental process as mentors work with new employees.

Ask Protégés to Critique Your Feedback Sessions

There is a sense in which getting corrective feedback from a mentor is like being a child receiving a scolding from the teacher. This is not the relationship that is wanted between a mentor and protégé. The desired relationship between mentor and protégé is adult-to-adult, not adult-to-child. Of course, one adult is more experienced and a mentor, but both are adult colleagues on the same team. To make sure the adult-to-adult relationship is maintained, mentors should ask protégés to critique feedback sessions. This will show protégés that they are, in fact, colleagues whose opinions are valid and wanted.

Asking protégés to critique feedback sessions can also reveal approaches, attitudes, and strategies the mentor needs to change, improve, or just drop altogether. A one-on-one critique between a mentor and protégé will send the message that you and I are in this together and that I need your help just as you need mine. Let's work together. This is an adult-to-adult message, and it is the right one to send to new employees.

Encouraging Protégés to Internalize the Safety-First Corporate Culture

An important responsibility of the mentor is encouraging protégés to adopt and internalize the organization's safety-first corporate culture. Generally speaking, this is accomplished by modeling and reinforcing appropriate attitudes and practices, while pointing out and correcting inappropriate ones. Before proceeding further, a clarification of terms is in order. The term *reinforcing,* as used in the current context, means taking appropriate action to encourage the behavior in question, such as providing positive feedback. The term *correcting,* as used here, means taking appropriate action to change inappropriate behavior.

An effective way to encourage safety-first attitudes and practices in new employees is to provide reinforcement when those attitudes and practices are observed. There are many ways protégés' actions can be reinforced by mentors. Some of the best forms of positive reinforcement cost nothing. For example, a public pat on the back and a few encouraging words that say "good job" are among the most effective forms of positive reinforcement. The only limits to what can be given as rewards for a job well done or for exemplifying the safety-first corporate culture are those imposed by your organization.

Consequently, before assigning experienced employees to serve as mentors, it is important for organizations to decide what mentors can and cannot give to reinforce safety-first attitudes and practices. Once again, this means the safety committee will have to work with the human resources department and, in turn, secure the approval of higher management for its recommendations. It is also important to provide mentors with a budget for any rewards that do have a cost associated with them. For example, if one of the rewards mentors can give their protégés is taking them to lunch as a way of saying "well done," the mentors should have a budget to cover the associated costs.

An excellent tool for developing a menu of potential reinforcement rewards for new employees is a book by Bob Nelson entitled *1001 Ways to Reward Employees*. This book contains examples of various kinds of nonmonetary and monetary reinforcement techniques that have been used effectively by organizations to encourage the desired attitudes and practices in employees. For example, here are just a few no-cost ways that are recommended in Nelson's book to recognize employees:[1]

- Post a "good-job" note on the employee's desk or workstation.
- Volunteer to answer the protégé's telephone and cover for her so she can leave work early one day.
- Have the organization's CEO send a handwritten "well-done" note to the protégé.

Simple strategies such as these can be surprisingly effective. However, a word of caution is in order here. Even the most appealing reinforcement rewards will fail to produce the desired results unless they are properly handled. What follows are strategies for maximizing the effectiveness of rewards given to protégés in an attempt to reinforce positive safety-first attitudes and practices.

Make Reinforcement Rewards as Immediate as Possible

A concept from the field of quality management known as "just-in-time" applies here. This concept has to do with providing things just at the time when they are needed and will do the most good. It is applied in a variety of ways ranging from inventory practices to the scheduling of training. It can also be applied in a mentoring situation to the provision of reinforcement rewards, and it should be.

Just-in-time reinforcement rewards are those given as close as possible to the practice or attitude being rewarded. A management principle to remember here is that the closer the reward to the behavior, the greater the effect of the reward. For example, if a new employee is observed applying all of the proper techniques when manually lifting boxes, a public compliment from the mentor that is given in the moment will have a much greater effect than the same compliment written on his performance appraisal months later. This does not mean that this employee should not be commended during his performance appraisal. Rather, it means that he should be commended publicly right now and then again during the performance appraisal.

Give the Reinforcement Reward Publicly

An important rule of thumb to remember when providing employee recognition is "praise in public and correct in private." The value of a publicly given reward is the attention and esteem it brings to protégés from colleagues, superiors, and others. This is why military personnel respond so positively to the medals they earn for valor in combat and why Olympic champions respond so positively to the bronze, silver, and gold medals they win. The decorations themselves are really just attractive pieces of medal and ribbon that have little monetary value, but the attention, prestige, public acclaim, and esteem they give their recipients are priceless.

Make Sure the Reinforcement Reward Fits the Behavior

Few organizations do as good a job as the military in making sure that the reward fits the behavior being rewarded. They do this by stratifying rewards in clearly defined levels. For example, military combat decorations for valor in combat in order of importance from lowest to highest are the Bronze Star, Silver Star, Distinguished Service Cross (Distinguished Flying Cross for the air force and Navy Cross for the navy and Marines), and the Medal of Honor. The Purple Heart stands alone as the medal awarded to recognize those who are wounded in action against the enemy. Acts of valor that would lead to the award of the Bronze Star medal, although commendable, would be much less than those that would lead to the award of the Medal of Honor, which is as it should be.

Organizations attempting to use rewards to reinforce appropriate safety-first attitudes and practices should follow the lead of the military and fit the reward to the deed. People have an innate sense of "fit" when it comes to rewards. Throwing a banquet for an employee who arrives at work on time for his first two weeks would be going overboard and everyone would know it, including the employee. On the other hand, just giving a pat on the back to a new employee whose effective handling of a hazardous situation saved the lives of several employees would be a serious understatement. There is an old saying about the punishment fitting the crime. Similarly, rewards should fit the behavior.

Use a Variety of Reinforcement Rewards

One thing many people have in common is an inherent need for variety. This is especially true of your organization's younger employees, who are accustomed to 150 or more television channels and an almost endless list of options in every other aspect of their lives. Although for some people there is comfort in sameness, for most people a little variety is more palatable. For example, do you like to eat the same thing at every meal or have a little variety in your diet?

For leisure and recreation do you like to do the same thing all the time, or would you rather do something different every now and then? Most people respond better to an appropriate level of variety than to the monotony of sameness. This is certainly the case with reinforcement rewards. Even the best reward will lose its value if given over and over again with no change. With reinforcement rewards, it is better to mix things up and have a little variety.

Assigning a mentor at the end of your organization's orientation can be an effective strategy for facilitating the process that leads to the development of a safety-first attitude in new employees. However, potential mentors should be vetted carefully. They must be employees who look and talk the part well, but also who actually believe in the organization's safety-first cultural imperatives. A good mentor is invaluable, but a bad mentor is worse than no mentor at all.

SAFETY-FIRST CORPORATE PROFILE
Eaton Corporation

Eaton Corporation is a diversified manufacturer of electrical systems and components for (1) power quality, distribution, and control; (2) fluid power systems and services for industrial, mobile, and aircraft equipment; (3) intelligent truck drivetrain systems for safety and fuel economy; and (4) automotive engine air management systems, powertrain solutions, and specialty controls for performance, fuel economy, and safety. Eaton employs more than 60,000 personnel in 125 countries.

Eaton exemplifies the most fundamental characteristic of organizations that are committed to a safety-first corporate culture. The company has a set of core values that guide its operations and decision-making process. Eaton is so committed to these values—one of which is workplace safety—that the company will lose business before it will compromise its values.

Eaton's commitment to workplace safety is set forth in its statement of corporate responsibility, which reads as follows: "Creating safe workplaces and conserving world resources." This is a simple, brief, but powerful statement of commitment. All key personnel understand the full ramifications of this commitment and their need to be mentors and provide mentors to ensure that all employees worldwide share the commitment. This is why *Ethisphere Magazine* named Eaton one of the world's most ethical companies.

Source: http://www.eaton.com and http://ethisphere.com/2007-worlds-most-ethical-companies/

COLLEAGUE-TO-COLLEAGUE DISCUSSION CASES

CASE 1: We Need Mentors in This Company

Jackson Edwards, safety manager at ABC Communications, Inc., is pleased with the progress the company has made in establishing a safety-first corporate culture. The new orientation program, developed jointly by ABC's human resources department and safety committee, is excellent. The emphasis of the program is now less about boilerplate facts and forms and more about beginning the process of instilling a safety-first attitude in new employees. But lately, Edwards has noticed a problem.

In talking with new employees who have recently completed ABC's new orientation program, Edwards has found that much of what was presented has been forgotten or was misunderstood. He is afraid that much of the initial value of the new orientation is being quickly lost in the weeks following its completion. One of the new employees he talked with described the orientation as analogous to trying to take a drink of water out of a fire hose.

After studying the "information overload" problem himself and discussing it with his colleagues on ABC's safety committee, Edwards is convinced that the company needs to provide mentors for new employees on the day they complete their orientation. However, this would be a major undertaking, and Edwards is not sure how to go about making this happen or even how to convince higher management that it is necessary.

Discussion Questions

1. Have you or any of your colleagues ever confronted a situation in which an organization's orientation program resulted in information overload for new employees? If so, discuss what kinds of problems this caused and what, if anything, the organization did about it.

2. If Jackson Edwards asked you and your colleagues how to go about selling the idea of providing mentors for new employees to higher management and how to actually implement a mentoring program once approval is given, what would your group advise?

Case 2: The Bad Mentor

Casey Weldman had appeared, at least on the surface, to be an excellent candidate to become an outstanding mentor. He was good at his job and seemed to exemplify his employer's cultural imperatives as they related to safety and other important values. Weldman had completed his company's mentor training program with high marks and was highly recommended by the trainer. This is why the current situation is so difficult to understand.

In spite of Weldman's outward appearances, the employees he mentors invariably turn out to be some of the company's worst and most consistent violators of safety practices. When closely supervised, they follow the rules and display a positive attitude toward safety. But when not closely supervised, they are prone to take shortcuts and ignore safety procedures. In addition, they respond to peer pressure to work safely with a contemptuous and condescending attitude.

Discussion Questions

1. Have you or any of your colleagues ever worked with people who worked safely when being observed, but ignored safety when left alone? What kinds of problems did this cause?

2. If you were the safety manager in the company in this case, how would you handle this situation?

Key Terms and Concepts

Before leaving this chapter, make sure you understand the following key terms and concepts and can accurately explain them to people who are not safety professionals.

Cultural/safety-first purpose of the mentor

Rookie jitters

Safety-first component of the orientation

Active, patient listener

Work to establish trust

Be empathetic

Be a consistent, positive role model

Show sincere interest in protégés

Be encouraging—begin with something positive

Think about what you are going to say before saying it

Give feedback frequently

Find protégés doing things right, and reinforce those behaviors

Make sure protégés understand what you are telling them and why

Be specific and give examples

Ask protégés to critique your feedback sessions

Make rewards as immediate as possible

Make the reward publicly

Make sure the reward fits the behavior

Use a variety of different rewards

Review Questions

Before leaving this chapter, make sure you can accurately and comprehensively, but succinctly, answer the following review questions:

1. Explain the cultural/safety-first purpose of the mentor.
2. List and explain the mentoring strategies for having a positive influence on protégés.
3. List and explain the best practices for giving feedback to protégés.
4. Explain how your organization can encourage protégés to internalize the safety-first corporate culture.

Application Project

Establishing an organization-wide mentoring program is a major undertaking that will require the safety committee to work with your organization's human resources department and to secure approval from higher management. Before approaching the human resources department, it is a good idea for the safety committee to have a plan for the development of a mentoring program and a "strawman" for the required training program.

Develop a plan for your organization. The plan should include at least the following: (1) a brief but comprehensive rationale for the program; (2) a proposed selection process for choosing mentors; (3) a proposed method for rewarding mentors for their time, effort, and expertise; and (4) the "strawman" for the mentoring training program.

Endnote

1. Bob Nelson, *1001 Ways to Reward Employees* (New York: Workman Publishing Company, 2005), 26.

Chapter 6

Train Personnel
in the Expected
Safety-First Attitudes
and Practices

Major Topics

- Rationale for training in safety-first organizations
- OSHA's training requirements
- Training assistance and resources available from OSHA
- Safety-first attitudinal training
- Safety professionals as trainers
- Training supervisors to carry out their safety-first role

Training has long been recognized as a critical requirement in organizations that are serious about workplace safety and health. Organizations that expect their personnel to work safely must be willing to teach them how. In other words, they must be willing to provide effective ongoing training. In organizations that are attempting to establish a safety-first corporate culture, training is even more important. This chapter explains how to use ongoing training to help instill safety-first attitudes and practices in your organization's personnel.

Rationale for Training in Safety-First Organizations

The reasons for providing comprehensive, ongoing safety training in today's hypercompetitive business environment are similar to those for establishing a safety-first corporate culture: (1) compliance, (2) ethics, and (3) competitiveness. In addition, just as each of the reasons for establishing a safety-first corporate culture has an associated cost rationale, so does each of the reasons for providing safety training. The cost rationale relating to safety training is simple: failing to train personnel to work safely can create circumstances in which your organization's financial resources are continually depleted by the costs of lawsuits, medical bills, insurance, and all of the other direct and indirect costs associated with accidents and injuries. In short, training is a critical component of the model for establishing a safety-first corporate culture.

Compliance Rationale for Safety Training

The OSH Act (Occupational Safety and Health Act) mandates that employers provide safety and health training. The act requires, among other things, the following:

- Education and training programs for employees
- Provision of information about all hazards to which employees will be exposed on the job
- Provision of information about the symptoms of exposure to toxic chemicals and other substances that might be present in the workplace
- Provision of information about emergency treatment procedures

To comply with the OSH Act, employers must satisfy these broad requirements and the more specific requirements relating to them. OSHA's specific requirements relating to training are set forth in Title 29—Code of Federal Regulations, as follows:

- General industry: Part 1910
- Maritime industry: Part 1915–1918
- Construction industry: Part 1926
- Agriculture industry: Part 1928

Ethics Rationale for Safety Training

The ethics-related reasons for providing safety training for your organization's personnel are similar to those for establishing a safety-first corporate culture (as explained in Chapter 1). Since organizations have an ethical obligation to provide a safe and healthy work environment for their personnel, it follows that they have a similar ethical obligation to provide appropriate training. Why? Part of providing a safe and healthy work environment is ensuring that your organization's personnel know how to work safely, that they are not—out of ignorance—safety hazards to themselves and their teammates. In other words, since personnel are a major component of the work environment, an organization cannot live up to its ethical obligation to provide a safe workplace unless its personnel know how to work safely and have positive attitudes about doing so.

Competitiveness Rationale for Safety Training

The competitiveness rationale for establishing a safety-first corporate culture is based on the need of organizations to achieve consistent peak performance and continual improvement in order to survive in today's hypercompetitive marketplace. To do what is necessary to achieve consistent peak performance and continual improvement, organizations must provide a safe and healthy work environment for their personnel—an environment that accommodates and encourages creative thinking, intense concentration on the work, and constant innovation. Recall from Chapter 1 that employees cannot focus on giving their best performance and continually improving it when they are constantly distracted by fear for their safety or resentment toward the organization because of accidents that injure their teammates. Nor can they think creatively and constantly innovate. A safe workplace provides the kind of environment necessary for peak performance and continual improvement.

Consequently, the competitiveness rationale for safety training logically follows the rationale for establishing a safety-first corporate culture. Organizations cannot provide a safe and healthy work environment if their personnel do not know how to work safely and, worse yet, do not care to. For your organization to have a safe and healthy work environment, its personnel must understand safe work practices and commit to using them—a commitment that is a natural by-product of a positive attitude toward safety; in other words, a safety-first attitude. One of the best ways to ensure that all personnel understand safe work practices and develop a safety-first attitude toward their work is to provide specialized training for them.

OSHA's Training Requirements

OSHA's training requirements, as shown in the previous section, are set forth in broad industrial categories: general, maritime, construction, and agriculture. Standards for the general industry category are contained in 29 CFR 1910. For example, here is a brief and partial list showing some of the training requirements contained in 29 CFR 1910:

- *PPE.* 29 CFR 1910.132(f)(2) requires that employees be able to demonstrate that they know how to use PPE as appropriate to their jobs.
- *Confined spaces.* 29 CFR 1910.146(g)(1) requires that employees who work as entrants, attendants, or entry supervisors be able to demonstrate that they have the understanding, knowledge, and skills necessary for the safe performance of their assigned duties.
- *Respiratory protection.* 29 CFR 1910.134(k)(1) requires that employees be able to demonstrate how to inspect, put on, remove, use, and check the seals of respirators.
- *Lockout/tagout.* 29 CFR 1910.147(c)(7)(i) requires that employees be able to demonstrate that they have the knowledge and skills required for the safe application, use, and removal of energy controls.

These are just a few of the many training requirements mandated by OSHA. Each requires that employees be able to demonstrate some type of understanding, knowledge, and/or skill relating to safe work practices. The best way to ensure that employees are able to demonstrate the required understanding, knowledge, and/or skill is by providing comprehensive, ongoing training.

Because safety professionals are familiar with OSHA's training requirements, it is not necessary to provide a comprehensive list of them in this chapter. However, it is a good idea to visit OSHA's website periodically at http://www.osha.gov and review the training requirements that apply to your organization. If you are a new safety professional, make sure that you familiarize yourself with all applicable training requirements right away.

Complying with OSHA's training requirements will satisfy your organization's training needs relating to safe work practices. This is important, but it is only half of the training that is needed in safety-first organizations. The other half is training aimed at developing a safety-first attitude. Attitudinal training is not mandated by OSHA. However, it is critical to the establishment and maintenance of a safety-first corporate culture. A recommended training program for instilling a safety-first attitude in employees is presented later in this chapter.

Training Assistance and Resources Available from OSHA

Experienced safety professionals are familiar with the training-related assistance and resources available from OSHA. Consequently, this section is for new safety professionals. OSHA's Directorate of Training and Education (DTE) develops, directs, oversees, and manages the agency's national training and educational policies and procedures. To ensure that OSHA's policies and procedures relating to training are effectively implemented, its DTE maintains several institutes, centers, and programs as follows:

- *OSHA Training Institute (OTI).* The institute provides safety and health training for federal and state compliance officers, state consultants, and personnel in various federal agencies. There are also provisions for training private-sector personnel.
- *OSHA Training Institute Education Centers.* The centers offer OTI's most frequently requested training programs for the private and public sectors at various locations across the United States.
- *OSHA Outreach Training Program.* The program provides training for individuals who want to become qualified as OSHA trainers in their organizations. Those who successfully complete an OSHA trainer course are authorized to teach groups of employees. For example, if a member of your organization's safety committee completes an OSHA trainer course, that individual may, in turn, provide OSHA training for employees of your organization.
- *OSHA Resource Center Loan Program.* OSHA's resource center loans training videos/DVDs to authorized trainers, including people who have completed OSHA's trainer program.

New safety professionals or experienced professionals who are unfamiliar with OSHA's training-related services and resources should spend some time learning about what is available to them and their organizations by visiting OSHA's website at http://www.osha.gov.

Safety-First Attitudinal Training

What makes a safety-first organization's safety training program different from those in other organizations is the attitudinal component. In addition to the training required by OSHA and other training related to safe work practices, safety-first organizations provide training aimed at instilling a positive attitude toward safety. The program I recommend and have used with success has the following components: (1) presentation and discussion of the definition of a safety-first corporate culture, (2) presentation and discussion of the rationale for adopting a safety-first corporate culture, (3) presentation and discussion of evidence of higher management's commitment to a safety-first corporate culture, and (4) presentation of group application activities structured around safety-related case studies. Each of these components of the attitudinal training is explained in the paragraphs that follow.

Safety-First Corporate Culture Definition Reviewed

In the Introduction to this book, a safety-first corporate culture was defined as follows:

> *A corporate culture in which the tacit assumptions, beliefs, values, attitudes, expectations, and behaviors that are widely shared and accepted in an organization support the establishment and maintenance of a safe and healthy work environment for all personnel and other stakeholders.*

In this first part of the safety-first attitudinal training, participants are given a copy of this definition and asked to read it. Then the instructor leads the group through a discussion of how this definition might apply in the workplace on a daily basis and what it means to each participant on a practical basis. This element of the training can typically be completed in 30 minutes or less.

Importance of a Safety-First Corporate Culture

In this part of the safety-first attitudinal training, the trainer lists on a flip chart or marker board the following reasons why a safety-first corporate culture is important: compliance, ethics, competition, and cost. Then, one at a time, each reason is explained. The explanations are those provided in Chapter 1. After each reason has been explained, the trainer leads the group through a discussion of the reason and why it is important. I recommend using examples of fines that have been levied against organizations for non-compliance, medical costs relating to specific accidents, and injuries and deaths that have occurred. The professional literature on safety is replete with examples of this nature. Using these kinds of examples is an excellent way to make the importance of a safety-first corporate culture real to participants. This part of the training can be completed in approximately 30 minutes.

Evidence of Higher Management's Commitment to Safety

In this part of the safety-first attitudinal training, the trainer gives each participant a copy of the following items: (1) the organization's strategic plan, (2) the organization's corporate safety policy, (3) the job description for his or her work, and (4) the applicable performance appraisal form. Each of these items contains an element relating to safety and is presented as evidence of higher management's commitment to a safety-first corporate culture. The trainer leads the group through a discussion of how the safety aspect of each item is evidence of higher management's commitment to safety. This part of the training can be completed in approximately 30 minutes.

Group Activities/Case Studies

At this point, the appropriate information relating to a safety-first corporate culture has been presented and discussed. The next step—a critical step that is often overlooked in training programs—involves applying what has been presented. An unalterable principle of learning is that real learning has not occurred until it can be properly applied. The group activities/case studies in this part of the training allow participants to apply what was presented in the previous parts.

SAFETY-FIRST FACT

Better Safety Equals Better Productivity and Less Turnover

According to a study conducted by the European Agency for Safety and Health at Work, there is a strong correlation between occupational safety and productivity. This study found that the higher an organization's workplace safety standards, the higher its productivity. In some cases, a good safety record can even be used to predict profitability. An additional and related benefit of high safety standards is less turnover, a major factor in determining an organization's productivity and profitability.

Source: Retrieved from http://osha.europa.eu/press_room/050121_CSR/newsarticle_view on March 22, 2008

The approach used in this part of the training involves several steps: (1) participants are divided into groups of three to five; (2) each group is given a set of case studies; (3) each group is asked to select a *recorder* and a *reporter* (the recorder makes notes for the group concerning how it thinks the questions that accompany the case should be answered, and the reporter explains the group's answers to the larger group); (4) each group reads the first case study and decides how its accompanying questions should be answered (groups may use all of the safety-related information presented to them up to this point as tools in answering the questions); (5) the reporter for each group stands up and explains how his or her group answered the questions that accompany the case; (6) the trainer leads the overall group through a discussion of the various responses from each individual group, irons out differences, and guides all participants toward a consensus response; and (7) the process is repeated for the rest of the case studies given to the groups (each group should complete at least two case studies). Each group devotes around 20–30 minutes to each case. The overall group discussion and consensus requires another 20–30 minutes.

The entire safety-first attitudinal training program, as set forth herein, can be completed in less than four hours. An important element of the training is the case studies. Case studies are developed by safety professionals. In fact, I have had a measure of success making development of the cases a project of organizations' safety committees. With experience, developing challenging, realistic cases studies is not difficult. After all, they are typically based on the experience of safety professionals. However, for those who have never developed a case study and are concerned about being able to, the following cases are provided as examples.

EXAMPLE CASE 1: Production versus Safety

John Brown is still a fairly new employee, having worked at ABC, Inc., for just six months. So far everything about his new job has gone well. That is, it had until earlier this week. Three days ago Brown's supervisor called a team meeting and explained that ABC, Inc., was behind schedule on an important project and that Brown's team was being pulled off of its current work to help get the lagging project back on schedule. The supervisor told Brown and his teammates that they were going to be watched to see how they responded to this crisis. "This is our big opportunity to show what we can do. If we can rescue the company on this project by getting it shipped on time, it will mean bonuses for all of us, and it will look good on our next performance appraisal."

Brown and his teammates responded positively to the challenge and, as soon as the work was delivered, they got right on it. However, after just two days it was clear that making the deadline would be a close call even with Brown's team helping out. In response, Brown's supervisor tells his team members, "I don't care what you have to do to make it happen. I want our part of this job finished on time." Apparently, Brown's teammates understood their supervisor's message because they are not just working longer and faster than normal, they are ignoring important safety precautions. There have already been several minor injuries and a couple of serious near misses. In spite of this, the supervisor is unrelenting in his pressure to speed up the production rate.

What Brown and his team are doing goes against everything they were taught during their orientation and in the mandatory safety training they recently completed. So far Brown is still following applicable safety procedures to the letter, but the pressure on him to cut corners is mounting. Just a few minutes ago his supervisor commented out loud so that others heard him: "What's wrong with you, Brown? Don't you want our team to get the performance bonus from this project?" Brown is uncomfortable ignoring safety procedures, but it is becoming clear that his teammates are turning out more work than he is in spite of the minor injuries some of them have sustained. He does not know what to do.

Discuss this case with your team members, and answer the following questions as a group: (1) Do you think ABC, Inc., has a safety-first corporate culture? (2) Do you think Brown should give in to the pressure to ignore safety procedures? (3) If you think Brown should resist the pressure to ignore safety procedures, how should he handle this situation?

EXAMPLE CASE 2: The Unsafe Manager

Sally Ryan is a member of XYZ Corporation's safety committee. Twice this week, Ryan has observed a certain manager who frequently visits the shop floor do so without donning the required safety glasses and hardhat. Other employees have noticed these safety violations too, but no one seems to be willing to confront the manager in question—a manager known to have a very short fuse with employees. Clearly this manager sets a bad example every time he violates a safety procedure established by the safety committee and approved by higher management. Ryan would like to take some action, but is unsure of what to do.

Discuss this case with your teammates, and answer the following questions as a group: (1) If you were an employee of XYZ Corporation, how would this manager's casual attitude toward safety procedures affect you? (2) If XYZ Corporation claims to be a safety-first company, what types of documentation should the safety committee have to show higher management's supposed commitment to safety? (3) If you were in Sally Ryan's place, how would you handle this situation?

Safety Professionals as Trainers

In organizations that maintain a safety-first corporate culture, safety professionals must be prepared to serve as in-house trainers. Training relating to safety-first work practices and attitudes is necessarily ongoing in safety-first organizations because new employees must be trained and experienced employees retrained as standards, procedures, technology, and circumstances change. To be an effective trainer, a safety professional should have certain characteristics, knowledge, and skills.

Characteristics of Effective Safety Trainers

Safety professionals who serve as in-house safety trainers for their organizations wear two hats. The first is that of the safety expert, who is knowledgeable of all standards, regulations, guidelines, procedures, and practices that apply in his or her organization. The second is that of the effective trainer, who, in addition to being a subject-matter expert, is able to effectively prepare, present, apply, and evaluate safety instruction. To effectively prepare, present, apply, and evaluate safety instruction, safety professionals should be knowledgeable about the fundamental principles of learning and the four-step teaching method.

Fundamental Principles of Learning

The principles of learning summarize what is known and widely accepted about how people learn. Trainers can do a better job of facilitating learning if they understand and apply the following principles:

- *People learn best when they are ready to learn.* You cannot make employees learn anything. However, you can show them why certain topics and practices are important so they will want to learn them. Consequently, it is a good idea to spend a little time at the beginning of every training session demonstrating in terms that are personal to participants why the topic in question is important. Letting employees know how they might benefit personally from the training in question will get them ready to learn. Safety is one of those topics that is easy to put into personal terms.

- *People learn more easily when what they are learning can be related to something they already know.* Build today's learning on what was learned yesterday and tomorrow's learning on what was learned today. Begin each new training session with a brief review of the previous one.

- *People learn best in a step-by-step manner.* This is an extension of the preceding principle. Learning should be organized into logically sequenced steps that proceed from the concrete to the abstract, the simple to the complex, the known to the unknown, and/or the big picture to the details.

- *People learn by doing.* This is probably the most important principle of learning for trainers to understand. Inexperienced trainers often confuse talking with teaching. Explanations are certainly an important part of the teaching and learning process. However, real learning occurs only when what has been explained and discussed is applied. To illustrate this point, consider the example of teaching a class to ride a bicycle. You can explain how it is done and even demonstrate the process, but until your students climb and start pedaling, they will not learn how to ride a bicycle. The meaning for safety trainers is evident: Don't just demonstrate correct safety procedures; make all employees demonstrate that they can apply them.

- *The sooner and more often people use what they are learning, the better they will remember and understand it.* How many things have you learned in your life that you no longer remember? High school algebra for example? People forget what they do not use. Trainers should keep this principle in mind when developing their lessons. It means that repetition and application should be built into the training. It also means training should be provided on a just-in-time basis.

- *Success in learning tends to stimulate additional learning.* This principle is similar to the management maxim that says "success breeds success." For trainers, it means dividing lessons into short enough segments so that participants can see progress (experience success), but not so short that they are not challenged. Effectively applying this principle requires finding the right balance between making lessons short enough, but not too short. This requires the intuition that comes from experience and is learned primarily through trial and error.

- *People who are learning need immediate and continual feedback.* Did you ever take a test and have to wait a week or more to get the results back? If so, that was probably a week later than you wanted them. People who are learning want to know how they are doing, and they want to know immediately and continually. Feedback from a trainer can be as simple as an approving nod, a pat on the back, or a comment such as "good job" or "that's almost right—let me show you again." Regardless of the form it takes, feedback should be given as soon and as often as possible throughout the learning process.

These principles should be applied before safety professionals even step into the training room. The process of applying these principles begins during the preparation step of the four-step teaching method and applies in all four steps. These steps are explained in the next section.

Four-Step Teaching Method

When wearing their "trainer hat," safety professionals are teachers, and teaching is about helping people learn. One of the most effective approaches for helping people learn is not new, gimmicky, innovative, or high tech in nature. In fact, it consists of the four building blocks that form the foundation of all good teaching. Innovative teaching techniques and instructional technologies simply allow trainers to build on the foundation established by the four-step teaching method. A description of each of these steps follows:

- *Preparation.* This step encompasses all tasks that are necessary to get participants prepared to learn, trainers prepared to teach, and facilities prepared to accommodate the process. Preparing people to learn means motivating them to want to learn (recall the first principle of learning from the previous section). This is easily done in safety training since there are immediate tangible benefits to working safely that can be demonstrated in terms that are personal to learners. Personal preparation (trainers preparing themselves to teach) involves preparing lessons and all of the material and instructional aids needed to support them. Preparing the facility involves arranging the room or workstation for both function and comfort and checking all instructional technologies, equipment, tools, and aids to make sure they are available, accessible, and usable.

- *Presentation.* This step involves presenting the material participants are supposed to learn. It might involve giving a demonstration, presenting a lecture, conducting a question-and-answer session, helping participants interact with an online program, or assisting participants who are working their way through self-paced training materials. Strategies that will enhance the effectiveness of presentations include the following: begin dramatically (show the result of failing to follow safety precautions), be brief, be organized, keep it simple, maintain your sense of humor, be patient, take charge, be sincere, and use real-life examples and stories from your experience whenever possible.

- *Application.* This step involves giving participants opportunities to apply what they are learning. For example, if you have just demonstrated proper lifting techniques, require participants to lift several objects while you observe and correct them where necessary. Continue the application step until all participants can demonstrate the proper techniques. Even if what you are teaching is strictly "bookwork," such as a new OSHA standard, require participants to demonstrate their understanding by explaining—verbally or in writing—what the standard means in practical terms.

- *Evaluation.* This step involves determining the extent to which the planned learning has taken place. In safety training, evaluation does not need to be a complicated process. In fact, it can be as simple as having participants demonstrate that they can properly do what you have demonstrated or effectively explain what you have presented. Safety training is not like a history class; a "C" is not sufficient. If participants are learning safe work practices, they must be able to demonstrate 100 percent proficiency before completing the training.

The Principles of Learning in the Four-Step Teaching Method

Various principles of learning presented earlier in this chapter can be applied beginning in the first step of the four-step teaching method and, in some cases, in all of the steps. This section explains how trainers can apply the principles of learning as they go through each step in the four-step teaching method.

Step 1: Preparation   When preparing the materials, facility, and participants, all of the principles of learning come into play. Since *people learn best when they are ready to learn*, trainers should be prepared to demonstrate in terms that are personal to participants why it is important for them to learn the subject matter in question. For example, I once attended a short class on PPE. The first topic presented was hardhats, which we all chaffed at having to wear (they messed up our 1960s hairdos). The trainer opened the class by placing a watermelon on a table. When he had our attention, another trainer standing high up on a ladder dropped a wrench that fell about 15 feet and shattered the watermelon right in front of our eyes.

Then, without saying a word, the trainer placed a hardhat on the table. Once again his colleague dropped a wrench, but this time it hit the hat hard and simply bounced off. The trainer then turned to the class and said, "Who can tell me why it is important to wear a hardhat in this facility?" This was a dramatic and effective way to motivate the class to pay attention and to understand why we needed to wear our hardhats.

Because *people learn more easily when what they are learning can be related to something they already know*, part of preparation is identifying knowledge that all participants are likely to have in common. This is easy for lessons two, three, and so on, but can be difficult for lesson one. Lesson two can be related to lesson one, lesson three to lesson two, and so on. But finding a common knowledge base for the first lesson can be a challenge. Sometimes it cannot be done. When this is the case, press on. This principle will come into play in subsequent lessons.

Because *people learn best in a step-by-step manner*, it is important to plan your lessons so they can be presented in this way. Teaching should proceed from the simple to the complex, concrete to the abstract, known to the unknown, and big picture to the details. For example, if presenting a lesson on the safe use of portable hand tools according to OSHA standard (29 CFR 1910.242) to a group of new employees, who might not be familiar with OSHA, you might plan to proceed as follows: (1) an explanation of OSHA and why it is so important to know and follow its standards, (2) how the standards are organized and which apply most directly to their work situation, (3) the actual content of the standard, and (4) how the standard applies on a practical basis every day on the job. This type of planning would ensure that teaching proceeds from the big picture to the details.

Since *people learn by doing*, it is important for trainers to prepare application activities and opportunities for learners. This involves finding ways to let them apply what they are learning. In your high school mathematics days, this involved working practice problems after the teacher had presented a new concept. In safety training, it means putting what is being learned to use. If students are learning about ladder safety, demonstrate the concept and then follow up the demonstration by having them properly place and climb a ladder, and perform some type of work while on the ladder. In other words, plan to have participants DO what you are teaching them.

Since *the more often people use what they are learning, the better they will remember and understand it*, and prepare to deliver instruction as close to its on-the-job application as possible. Ideally, participants will complete their training on a given concept and begin using what they have learned that very day on the job. When too much time elapses between the training and its actual on-the-job application, people are prone to forget what they have learned.

Since *success in learning tends to stimulate additional learning*, prepare lessons so that participants are required to bite off an amount that is challenging, but not overwhelming.

Prepare your lessons so that participants can experience success right away. For example, assume you are preparing to teach a program on safety precautions for employees who will work in a cold environment. You plan to present material in the following areas: (1) proper warming procedures, (2) applicable first-aid procedures, (3) protective clothing and its proper use, (4) proper eating habits, (5) proper drinking habits, (6) recognizing the symptoms of cold stress and strain, and (7) safe work processes.

The "success principle" means that you would divide the course into several lessons, perhaps one lesson for each numbered item in this list. Then for each lesson, you would have preparation, presentation, application, and evaluation before moving to the next lesson. Just presenting the material on all seven items in one comprehensive lesson would be a sure way to overwhelm participants. Each lesson should contain enough information to challenge participants, but not enough to overwhelm them.

Since *people need immediate and continual feedback* when learning, part of your preparation should involve planning ways to continually let participants know how they are doing throughout the training program. Feedback does not have to be formal. Just observing participants, as they attempt to demonstrate their mastery of the concept being taught, and complimenting or correcting them as appropriate can be an effective approach to providing continual feedback. The key is to include plans for giving feedback during your preparation.

Step 2: Presentation   All of the principles of learning apply in this step. This is the step in which you put into practice the plans you made in the previous step relating to each principle of learning (e.g., the watermelon demonstration for getting participants ready to learn the importance of wearing hardhats, relating the new learning to something participants already know, presenting the lesson in a step-by-step manner, building in opportunities for application).

Step 3: Application   The principle that applies most directly in this step is *people learn by doing*.

Step 4: Evaluation   The principle that applies most directly in this step is *people need immediate and continual feedback*. During the preparation step, you planned ways to apply this principle (e.g., by observing and correcting). Remember that in safety training, participants need to master the proper work practice and all of the information presented. Just making a "C" is not sufficient. In the workplace, almost getting it right can cause an accident or incident.

Training Supervisors to Carry Out Their Safety-First Role

In safety-first organizations, supervisors play an important role. They represent the first line of defense in monitoring the daily work practices of employees and in identifying hazardous conditions. A supervisor can be the safety professional's best friend or worse enemy depending on his or her attitude toward workplace safety. Consequently, it is important to ensure that supervisors are on the team when it comes to safety and health.

To help ensure that they properly carry out their safety-first role in your organization, I recommend that all supervisors be required to complete a comprehensive training program in which they learn (1) the safety-first role of supervisors and (2) the specific skills that are needed by supervisors in safety-first organizations. In addition, it is important to ensure that the safety-first roles of supervisors appear in their job descriptions and performance appraisals.

Goals and Objectives of Safety-First Training for Supervisors

Safety-first training for supervisors has one broad goal and a number of related and more specific objectives. The broad goal of safety-first training for supervisors is as follows:

> *To equip supervisors with the knowledge, skills, and attitude to play a positive role in helping the organization establish and maintain a safety-first corporate culture.*

Specific objectives of safety-first training for supervisors include but are not limited to the following (your organization may choose to add organization-specific objectives):

1. Help supervisors understand the importance of displaying a positive, safety-first attitude at all times (includes the definition of and rationale for a safety-first corporate culture).

2. Help supervisors understand the importance of setting an example of safe work practices at all times.

3. Help supervisors understand how to orient new employees to the safety-first way of doing their jobs.

4. Show supervisors how to ensure that their direct reports receive the safety training they need on a continual basis.

5. Show supervisors how to monitor employee work practices and attitudes on a daily basis and enforce safety rules and regulations.

6. Help supervisors become proficient in hazard analysis and accident prevention.

7. Help supervisors become proficient in job safety analysis.

8. Help supervisors become proficient in assisting safety professionals in conducting accident investigations.

9. Help supervisors become proficient in assisting safety professionals in writing accident reports.

10. Show supervisors how to stay up to date on safety-related issues.

Each of these objectives represents the main topic for one session in a comprehensive safety-first training program for supervisors. In addition to completing training relating to these objectives, supervisors should also complete the same types of technical safety training their direct reports have to complete. This includes (as appropriate to the job in question) training topics such as the following:

- Ergonomic hazards
- Mechanical hazards and machine safeguarding

- Falling, impact, and acceleration hazards
- Lifting hazards
- Vision hazards
- Temperature hazards
- Pressure hazards
- Electrical hazards
- Fire hazards and life safety
- Industrial hygiene
- Confined space hazards
- Radiation hazards
- Noise hazards
- Vibration hazards
- Automation-related hazards
- Blood-borne pathogen-related hazards

The NSC (National Safety Council) is an excellent source of help for safety professionals who are responsible for developing and presenting safety training for supervisors in their organizations. Its "Basic Safety for Supervisors" course, with the addition of cultural material specific to your organization, can make an excellent training program for supervisors. This course consists of 14 sessions as follows:[1]

Session 1: Loss Control for Supervisors. This session covers accidents and incidents, areas of responsibility, cost of accidents, and an effective approach to occupational safety and health.

Session 2: Communications. This session covers the elements of communication, methods of communication, and effective listening.

Session 3: Human Relations. This session covers human relations concepts, leadership, workers with special problems, and drugs and alcohol.

Session 4: Employee Involvement in Safety. This session covers promoting safe worker attitudes, employee recognition, safety meetings, and off-the-job accident problems.

Session 5: Safety Training. This session covers new-employee indoctrinations, job safety analysis (JSA), job instruction training (JIT), and other methods of instruction.

Session 6: Industrial Hygiene and Noise Control. This session covers general concepts, chemical agents, physical agents, temperature extremes, atmospheric pressures, ergonomics, biological stresses, threshold limit values (TLVs), and controls.

Session 7: Accident Investigation. This session covers finding causes of accidents, emergency procedures, effective use of witnesses, and accident reports.

Session 8: Safety Inspections. This session covers formal inspections, inspection planning and checklists, inspection of work practices, frequency of inspections, recording hazards, and follow-up actions.

Session 9: Personal Protective Equipment. This session covers controlling hazards; overcoming objections of employees to using PPE; protecting the head, eyes, and ears; respiratory protective equipment; safety belts and harnesses; protecting against radiation; safe work clothing; and protecting the hands, arms, legs, and feet.

Session 10: Materials Handling and Storage. This session covers materials handling problems; materials handling equipment; ropes, chains, and slings; and material storage.

Session 11: Machine Safeguarding. This session covers principles of safeguarding, safeguard design, safeguarding mechanisms, safeguard types, and safeguard maintenance.

Session 12: Hand Tools and Portable Power Tools. This session covers safe work practices, use of hand tools, use of portable power tools, and maintenance/repair of tools.

Session 13: Electrical Safety. This session covers a review of electrical fundamentals, branch circuits and grounding concepts, plug-and-cord-connected equipment, branch circuit and equipment testing methods, ground fault circuit interrupters, hazardous locations, common electrical deficiencies, safeguards for home appliances, and safety program policies and procedures.

Session 14: Fire Safety. This session covers basic principles; causes of fires; fire-safe housekeeping; alarms, equipment, and evacuation; and a review of the supervisor's fire job.

SAFETY-FIRST CORPORATE PROFILE
Pearson

Pearson is an international media company that specializes in education, business information, and consumer publishing. Pearson employs more than 30,000 personnel in 60 countries and is committed to the safety and health of each employee. To this end, Pearson has developed and deployed a comprehensive safety policy that contains several specific goals, one of which relates directly to training. The following is the company's safety-related training goal:

"Train, support, inform, and encourage employees to carry out their work safely."

To accomplish this goal, Pearson provides a comprehensive safety training program for its employees, supervisors, and managers that equips them to (1) do their part to ensure a safe and healthy work environment; (2) operate equipment and physical plants safely; (3) safely handle, store, and transport materials and substances; (4) respond promptly and effectively to emergencies; and (5) do their part to provide effective arrangement for dealing with work-related illnesses and injuries.

Source: http://www.pearson.com/index.cfm?pageid=148

This course may be more comprehensive than some organizations will need, but it can be easily modified onsite to fit the local exigencies of your organization. By adding organization-specific material relating to a safety-first corporate culture to the NSC's curriculum, you can create an excellent safety training program for supervisors.

COLLEAGUE-TO-COLLEAGUE DISCUSSION CASES

CASE 1: Why Do We Need All of This Safety Training?

Mike Simpson, safety engineer for ChemCo, Inc., chairs the company's safety committee. The committee has been working for several months to implement a plan for establishing a safety-first corporate culture at ChemCo. Overall, Simpson is pleased with the company's progress. However, at this morning's meeting of the safety committee, he ran into a snag. Mack Cramer, a supervisor in ChemCo's processing plant and a member of the safety committee, took exception to the proposed training program for supervisors.

After looking over the proposed curriculum for the required safety training for supervisors, Cramer asked the committee, "Why do we need all of this training?" He claimed to oppose mandatory training for supervisors because it would take them away from their normal duties for too long. The committee debated the issue back and forth for more than an hour, but when the meeting broke up, Cramer was still not satisfied. This has Simpson worried because Cramer is very influential with his fellow supervisors.

Discussion Questions

1. Have you or any of your colleagues ever had to deal with a supervisor(s) who was opposed to completing safety training? If so, what kinds of problems did this cause, and how were the problems dealt with?

2. If Mike Simpson asked you and your colleagues how to go about dealing with this situation, what would you tell him?

CASE 2: I'm a New Safety Manager Who Feels Lost

The young man sitting across the table from Margie Fellon appeared uneasy. They were both attending a conference sponsored by the NSC and wound up sitting at the same table for lunch. Fellon had been a safety professional for more than 20 years, so she could spot a "rookie" when she saw one. Fellon asked the young man his name and where he worked. He introduced himself as Donald Gilliam, told Fellon where he worked, and then said, "I'm a new safety manager who feels lost." Remembering her time as a brand new safety professional, Fellon could sympathize with Gilliam.

"Is there anything in particular that is causing you problems?" Gilliam, who was a recent college graduate with no work experience, responded that he felt overwhelmed by all he needed to learn. "But my biggest problem right now is that I have to develop a safety training program for the supervisors in my company, and I don't even know where to start." Gilliam then explained that since this was his first big assignment, he wanted it to go well. He needed to develop a course that would impress his boss and have credibility with supervisors.

Discussion Questions

1. Can you or any of your colleagues recall your first major assignment as a "rookie" safety professional? What was the assignment? Did you need help completing it? How did you handle the situation?

2. Put yourself in Margie Fellon's place. What types of assistance would you guide Gilliam to? What would you tell him about serving as a safety trainer for supervisors?

Key Terms and Concepts

Before leaving this chapter, make sure you understand the following key terms and concepts and can accurately explain them to people who are not safety professionals.

Compliance rationale for safety training	Characteristics of effective safety trainers
Ethics rationale for safety training	Fundamental principles of learning
Competitiveness rational for safety training	Four-step teaching method
OSHA's training requirements	Preparation
Training assistance and resources available from OSHA	Presentation
	Application
Safety-first attitudinal training	Evaluation

Review Questions

Before leaving this chapter, make sure you can accurately and comprehensively, but succinctly, answer the following review questions.

1. Explain the various elements of the rationale for providing safety training in safety-first organizations.

2. Specify where you would find OSHA's training requirements for your organization.

3. Summarize the training assistance and resources available from OSHA.

4. Describe the attitudinal aspects of safety training that should be provided in safety-first organizations.

5. Explain what you and your fellow safety professionals need to know at a minimum to be effective safety trainers.

6. Summarize the curriculum for a comprehensive safety training program for supervisors (do not forget the attitudinal component).

Application Project

Establishing a safety training program for supervisors is a major undertaking. Assume that the safety committee in you organization has agreed that such a program should be developed and that there is sufficient support from higher management. Develop a comprehensive plan for providing the training in your organization. Your plan should include the following elements:

1. List of resources and materials available from OSHA and the NSC.

2. Comprehensive course outline for a course for employees (including an attitudinal component).

3. Comprehensive course outline for a course for supervisors (including an attitudinal component).

4. Comprehensive summary of the facility, equipment, and materials that will be needed to implement your training plan.

5. Budget for your training plan.

Endnote

1. Retrieved April 7, 2008, from http://www.nsc.org.

Chapter 7

Make Safety Part of Team Building

Major Topics

Safety and
- the responsibilities of team members
- a teamwork-supportive work environment
- team skills
- the team charter
- team rewards
- team accountability

There is a natural and complementary link between safety and teamwork. One of the objectives of teamwork should be to help organizations promote safety-positive attitudes and work practices. One of the objectives of safety should be to help organizations maintain a work environment in which teamwork can flourish. Consequently, in safety-first organizations safety is a fully integrated component of teamwork and a team-building strategy. This chapter explains how organizations can make safety part of team building and teamwork.

Safety and the Responsibilities of Team Members

In teams, the members have responsibilities to the team and to each other. The most important responsibilities of team members in organizations are as follows:

- Participate fully, actively, and willingly and with a positive attitude in all team activities and assignments.
- Be punctual and regular in attendance for all team activities.
- Be open, honest, and frank with fellow team members.

- Be mutually supportive of team members—help others perform better.
- Make a concerted effort to get along with fellow team members.
- Be a good listener to other team members.
- Be open to the ideas of others.
- Look out for the health, safety, and well-being of fellow team members.

You will notice that the last responsibility on the list pertains directly to safety. However, there is also a safety-related aspect to every one of the responsibilities on the list, as explained in the paragraphs that follow. If this interrelatedness between teamwork and safety is understood, organizations can kill two birds with one stone. While building teams, they can also be building a safety-first corporate culture. The key to gaining this double benefit is for organizations' safety committees and safety professionals to work closely with managers and supervisors to ensure that safety is included in team building in the ways explained in the following paragraphs.

Participate Fully, Actively, and Willingly and with a Positive Attitude

Team members are expected to participate fully, actively, and willingly and with a positive attitude in all team activities and assignments. One of those activities, from time to time, will be safety training. Another consists of the everyday job tasks that must be performed. Individuals who maintain a positive attitude while fully, actively, and willingly completing periodic safety training and fully, actively, and willingly applying what they learn in the workplace make better team members. As supervisors encourage their team members to participate fully, actively, and willingly and with a positive attitude in all team activities and assignments, they need to stress that these activities and assignments include those relating to safety.

Be Punctual and Regular in Attendance

Punctuality and regularity in attendance are linked directly to workplace safety. Team members who come late to safety training or who miss it altogether put their teammates at risk because of what they do not know as a result of missing part of the safety training. Further, team members who arrive late for work or miss work altogether sometimes force their teammates to perform tasks shorthanded, a sure way to increase the likelihood of an accident or incident. When discussing punctuality and regular attendance with their team members, supervisors should explain how poor punctuality and attendance can put the health and safety of their teammates at risk.

Be Open, Honest, and Frank with Fellow Team Members

To challenge team members who are taking dangerous shortcuts and ignoring applicable safety precautions, people in teams have to be willing to be open, honest, and frank with each other. They have to be willing to say, "What you are doing is dangerous, and it needs to stop right now." This is not easy. At best, being open, honest, and frank about safety precautions can make people in teams uncomfortable. At worst, it can cause conflict. However, one of the benefits of teamwork, when it is done well, is that it engages peer pressure in support of safety-first attitudes and practices. When explaining to their team members why openness and honesty are so important to effective teamwork, supervisors should also explain how openness and honesty can affect the safety and health of team members.

Be Mutually Supportive of Team Members

In well-functioning teams, people support each other—they back each other up, pitch in to assist where and when necessary, and generally do what is necessary to help each other perform better. One of the ways people in teams support each other is in performing work tasks safely. For example, team members will do such things as (1) pitch in to help a teammate who is trying to lift a heavy object, (2) check the personal protective gear of teammates to ensure that everything works properly, (3) stand by outside of a confined space that a teammate has entered, (4) call a teammate's attention to the fact that a piece of equipment is tagged out, or (5) provide the support necessary to ensure that teammates are able to perform job tasks safely. As supervisors explain the performance-related benefits of mutual support to their team members, they should also explain how it can improve the safety of team members.

Make a Concerted Effort to Get Along with Fellow Team Members

Team members who make a concerted effort to work well with their teammates make themselves approachable. When team members need help to perform a task safely, they are more likely to ask for it from teammates they consider approachable. On the other hand, they will be less likely to ask for help, no matter how badly they might need it, from someone they do not get along with—someone who is unapproachable. As supervisors explain how getting along in teams can improve the team's performance, they should also explain how it can improve safety.

Be a Good Listener to Other Team Members

Team members who are good listeners are more likely to be approached by teammates who are concerned about hazardous conditions, unsafe procedures, or safety violations. A team member might have a safety-related concern and need someone to discuss it with, before pointing it out to the supervisor or before even deciding whether it needs to be pointed out. In addition, good listeners listen not just with their ears, but also with their eyes; they observe their teammates and the work environment. Consequently, good listeners are more likely than poor listeners to hear about or notice unsafe conditions in time to prevent an accident. Supervisors need to stress this point when they explain the importance of effective listening to their team members.

Be Open to the Ideas of Others

Team members who are open to the ideas of others are more likely to be approached by teammates who have recommendations for a better and/or safer way to perform a given job task. This is important from the perspectives of both continual improvement and safety. To improve their performance continually, teams must have members who are always looking for better ways to get the job done. To improve their safety record continually, teams must have members who are always looking for safer ways to get the job done. Team members who are open to the ideas of others encourage their teammates—by their openness—to constantly look for better ways and to share their thoughts rather than keeping good ideas to themselves. Supervisors should explain how being open, honest, and frank can improve both performance and safety.

Look Out for the Health, Safety, and Well-Being of Fellow Team Members

One of the benefits of effective teamwork is that team members grow closer and closer over time. As this happens, individual team members become increasingly concerned about the health, safety, and well-being of their teammates. They become protective of each other. When this happens, individual team members are less likely to take dangerous shortcuts, overlook hazardous conditions, or make ill-advised decisions that could lead to disaster. When team members look out for the health, safety, and well-being of each other, it is easier for organizations to maintain safe work practices and safety-first attitudes among their personnel. This is an important point supervisors should make to their team members.

Safety and a Teamwork-Supportive Work Environment

When organizations establish a teamwork-supportive work environment, they take a major step toward establishing a safety-first work environment. A teamwork-supportive work environment has the following characteristics:

- Open communication
- Constructive, positive interaction
- Mutually supportive approach to work
- Respectful interaction

Each of these characteristics has a safety-related aspect. It is important for your organization's safety committee and safety professionals to make sure that supervisors understand the safety aspects of these characteristics, and how to make them part of the team-building process.

Open Communication

Teamwork thrives in an environment where there is open communication between and among all stakeholders. In such an environment, there is effective, open communication between and among (1) individual team members, (2) team members and their supervisor, and (3) management and teams. When open communication exists between and among all stakeholders, team members feel free to (1) identify problems and issues relating to both performance and safety and point them out to their supervisors, (2) make recommendations to their supervisors about ways to improve performance and safety, and (3) discuss among themselves ways to improve performance and safety.

Constructive, Positive Interaction

Teamwork thrives in an environment where team-member-to-team-member and team-member-to-supervisor interaction is constructive and positive. By extension, so does safety. The purpose of interaction between and among personnel in high-performing organizations is to promote peak performance and continual improvement. To do this, interaction among employees, supervisors, and managers must be constructive and positive. People who respond to complaints, suggestions, and recommendations in a negative manner shut off communication. When this happens, the interaction that needs to occur if organizations are going to achieve peak performance and continual improvement is shut down, and both performance and safety suffer as a result.

Mutually Supportive Approach to Work

Mutual support is fundamental to teamwork. Teams function better when their members back each other up, pitch in to help one another, and look out for each other's safety. Consequently, when organizations establish a teamwork-supportive work environment, they also contribute to the establishment and maintenance of a safety-first corporate culture.

Respectful Interaction

Teamwork thrives in a setting where employees, supervisors, and managers treat each other with respect. Organizations that establish a teamwork-supportive work environment create such a setting. When people in organizations respect each other, they also respect each other's opinions. Consequently, they are open to suggestions and recommendations that are made in an effort to improve performance or safety. Consequently, energy and resources invested in establishing a teamwork-supportive work environment—an environment that, in turn, promotes respectful interaction—can gain a double return by improving both performance and safety.

Because of the safety-related benefits that accrue from having a teamwork-supportive work environment, your organization's safety committee and safety professionals need to be advocates of team building. They also need to be involved in the planning stages of team-building activities and training to ensure that the safety aspects of teamwork are built into the activities and training as well as the team charter (team charters are covered later in this chapter).

Safety and Team Skills

To be good team players, people need to have or develop several very specific skills. What I call "skills" in this section are usually referred to as characteristics or even character traits. I call them skills because they can be learned and continually improved over time. The following skills make people good team players:

- Honesty
- Selflessness
- Initiative
- Patience
- Resourcefulness
- Punctuality
- Tolerance
- Perseverance

Although most of these team-player skills appear to be character traits, it is important—from the perspective of team building and safety—to view them as skills. While it is true that some people might appear to be naturally more tolerant, honest, or patient than others, it is important to remember that these traits and all of the others listed above can be learned. Dishonest people can learn to be honest. Selfish people can learn to be selfless. Procrastinators can learn to take the initiative, and impatient people can learn to be patient. In fact, people have an almost unlimited capacity for learning when they see that what they need to learn is in their best interest.

The reason it is so important to approach the list of characteristics presented earlier in this section as skills is that human tendency is to think of them as characteristics people are born with, traits that they either have or do not have. This perspective provides fertile ground for making excuses (e.g., I just can't get to work on time—I'm not a punctual person; I just can't take the initiative—I am a procrastinator; or I just can't work on this any longer—I don't have any perseverance). These excuses are not valid. Like any skill, people can develop these and other team-player skills.

Consequently, a necessary part of team building is helping personnel understand how it is in their best interests to be honest, selfless, patient, and so on. This is an easy point to make when conducting team-building activities. The message that should be given to team members is this: (1) When teams perform better, our organization wins; (2) when our organization wins, we all win; (3) in a globally competitive marketplace, organizations that consistently win survive and thrive—a fact that allows them to provide job security and advancement opportunities for their personnel; and (4) in each of these cases, the opposite is also true.

In safety-first organizations, supervisors and others who participate in team-building activities and training understand that each of the team-player skills listed above has a safety aspect and approach team building accordingly. The safety aspects of each team-player skill are explained in the paragraphs that follow. The safety committee and safety professionals in your organization should work closely with supervisors, human resource personnel, and others who are responsible for developing and implementing team-building activities and training. They should ensure that the safety aspects of teamwork explained in the following paragraphs are given adequate attention during team-building activities and training.

Supervisors and other team leaders can help their personnel develop team-player skills by adopting the following multistep approach: (1) Let people know that these skills are expected of them, (2) be a consistent role model of what you expect of others, (3) communicate frequently about the expectations, (4) continually monitor the practices and attitudes of your personnel as they relate to team-player skills, (5) recognize and reward people who meet or exceed expectations, and (6) correct people who fall short of expectations.

Honesty

People learn to be honest by how they are raised and the type of environment they grow up in. Consequently, honesty can be taught. Supervisors teach their personnel to be honest by expecting honesty. They let their direct reports know that honesty is expected by making it part of the team charter (covered later in this chapter), talking about it with team members, monitoring the honesty of direct reports, recognizing and rewarding their personnel for honesty, and correcting personnel who exhibit a lack of honesty. Supervisors role-model honesty by setting a consistent example of being honest themselves.

Honesty in team members can have a powerful effect on safety-related attitudes and practices. Team members who are honest are less likely to cut corners and ignore safety procedures. They are also more likely to speak up when they see teammates doing these things. Honest team members are more likely to call potentially hazardous conditions to the attention of their supervisor and give an accurate assessment of what they see. Correspondingly, because they do not exaggerate or misinform, they are more likely to be listened to. This last point, especially as it relates to safety, should be made part of the team-building process.

Selflessness

If you adopt the perspective that people can be self-centered and self-interested, you will be right more often than you might like to admit. There are only so many people like Mother Teresa in the world. However, people can learn to be selfless or, at least, less self-centered. I recommend the multistep approach explained above for helping direct reports develop selflessness. Selflessness in team members can help ensure both safety-first attitudes and practices.

Selfless team members choose in all instances to do what is best for the team. When they understand that maintaining a safe and healthy workplace is best for the team, selfless team members will adopt a safety-first attitude and approach their work in a safe manner. They will also be more likely to insist that other team members do the same. Consequently, developing selflessness in team members is an excellent way to ensure that peer pressure works on behalf of safety. This point should be made part of the team-building process.

Initiative

Initiative is critical to teamwork and the maintenance of a safety-first corporate culture. People in teams should always be looking for ways to get the job done more effectively and efficiently. In other words, they should be committed to continual improvement.

SAFETY-FIRST FACT

Safety and Profitability Are Compatible

According to the United Kingdom's National Physical Laboratory, workplace safety and business profitability are fully compatible concepts. "The company goals of client satisfaction, ongoing business development and long term profitability are not in conflict with good safety practices. On the contrary, a good safety record is of great benefit to our business, whilst a safe and healthy work environment protects and encourages people, our most important asset."

Source: Retrieved from http://www.npl.co.uk/server.php?show=nav.383

Such a commitment requires initiative, another team-player skill that can be taught using the multistep approach already explained.

Initiative in team members also helps ensure a safe and healthy workplace. People in teams are constantly looking for ways to improve performance. This means they are also looking for circumstances or conditions that might inhibit performance. Hazardous conditions fall into this category. Consequently, part of team building should involve teaching personnel to take the initiative in identifying hazardous conditions and in helping find ways to eliminate or mitigate those conditions. Team members who take the initiative are more likely to do this, a point that should be emphasized as part of the team-building process.

Patience

Good team players are patient with their teammates. This is important because team members rely on each other as they work together to accomplish the team's mission. This can be a difficult team-player skill to develop because people do not like to have to rely on others. People feel more in control when they can rely on themselves and themselves only. However, a team cannot fulfill its mission if it is just a collection of self-reliant individuals, each going his or her own way, rather than a team of people pulling together on the same end of the rope. Like the other team-player skills, patience can be taught using the multistep approach recommended earlier in this chapter.

Patience can be an especially valuable asset for organizations attempting to establish and maintain a safety-first corporate culture. It comes into play at the team level when teammates patiently take the time to show each other how to safely perform job tasks. Sometimes it can take several demonstrations and corrective feedback that is repeated several times before a team member learns that the safe way is the right way. Team members who are willing to patiently but persistently demonstrate the safe way to their colleagues will promote a safety-first attitude much better than those who get frustrated with their teammates and simply give up. These safety-related points should be made part of the team-building process.

Resourcefulness

Resourceful people are those who will find a way to get the job done. People who work in teams constantly face the challenge of having to get their work done right and on time even when they are short of resources (e.g., time, personnel, material, supplies). Resourceful people never use a lack of resources as an excuse for failing to get the job done. Rather, they look at the lack of resources as just one more challenge. Resourcefulness can be taught using the multistep approach recommended earlier in this chapter.

Resourceful team members tend to be safer team members. Why? There are several reasons: (1) A safe and healthy work environment is itself an invaluable resource to a team; (2) team members are invaluable resources and should, therefore, be protected from accidents and injuries; (3) in addition to people, accidents, and incidents often damage such valuable resources as equipment, technology, machines, and more; and (4) accidents and injuries can sap an organization of its already limited resources. All of these safety-related points should be made part of the team-building process.

Punctuality

Because team members rely on each other to get their jobs done, punctuality is an important team-player skill. Team members who arrive late for work or who fail to complete their assignments on time undermine the work of their teammates. Mutual dependence is a fundamental aspect of teamwork. Consequently, punctuality is important in teams. Fortunately, punctuality is a team-player skill that can be taught using the multistep approach explained earlier in this chapter.

Punctual team members contribute in an important way to maintaining a safe and healthy work environment for their teammates. When team members are late, their teammates are forced to do more work with fewer resources. This is always a dangerous situation because it can lead to a hazardous work condition known as *overload*. People trying to do the work of two are more susceptible to accidents and injuries. This is an important point that should be made part of the team-building process.

Tolerance

Tolerance is an important team-player skill that has become even more important over the past two decades as the workplace has evolved into an increasingly diverse environment. Diversity in the workplace means that team members will work every day with people who are different from them in many ways, including age, race, gender, cultural heritage, nationality, political philosophy, native language, and religion. This means that to function effectively as part of a team, people must be tolerant of individual differences. However, it does not mean that teammates should tolerate poor attitudes or practices relating to safety. Tolerance can be taught using the multistep approach recommended earlier in this chapter.

Tolerant team members tend to be safer team members. There are several reasons for this. First, they tend to get along with their teammates better than intolerant people. People who get along are more likely to help each other perform potentially dangerous tasks and be mutually supportive in watching out for each other's safety and health. Second, people who are tolerant are more approachable to other team members who need to ask for help or who need advice concerning how to safely perform a given task. Finally, team members who are tolerant are more likely to be listened to when they attempt to apply peer pressure on behalf of safety. All of these points should be made part of the team-building process.

Perseverance

Perseverance is an important team-player skill because of the nature of work. Work is seldom easy (that is why it is called "work"). Sometimes the most successful teams are those staffed by people who are willing to just keep trying when others have given up. Perseverance is about continuing the effort even though you feel like quitting. Perseverance can be taught using the multistep approach explained earlier in this chapter.

Perseverance in team members promotes peak performance and continual improvement. It also promotes safety. Often accidents are caused by people who get tired or frustrated and start to take shortcuts, such as ignoring safety precautions. Team members who are willing to not just persevere in getting the job done but persevere in doing it safely make better team players and are more valuable employees. This point should be made part of the team-building process.

Safety and the Team Charter

A team charter is a document that shows employees the mission of their work team and the ground rules they will follow while working together to accomplish the mission. A team charter can contain many ground rules, including the following (notice that this list contains all of the people skills listed earlier that make a person a good team player as well as additional people skills):

- Honesty
- Dependability
- Selflessness
- Responsibility
- Mutual supportiveness
- Initiative
- Patience
- Punctuality
- Tolerance
- Resourcefulness
- Safety oriented

Ground rules are basic operating principles that define how team members are to interact with each other as they work together to accomplish the team's mission. Consequently, it is important that at least one of the team ground rules relates to safety.

Developing Team Charters

Team charters should be developed at the team level. However, if your organization does not yet have team charters or if it has them, but they do not contain a safety-related ground rule, the safety committee will have to take the lead in getting them developed or revised as appropriate. This means that members of the safety committee will have to (1) seek higher management's approval of the concept and (2) work with supervisors to develop team charters or revise existing charters as appropriate. I have found it helpful to bring the human resources department into the loop prior to seeking the approval of higher management. In fact, an organization-wide effort to develop team charters is best undertaken by the human resources department with assistance from the safety committee.

Developing the Team Mission Statement

Once the concept has been approved, the next step is for human resources personnel to work with supervisors organization-wide to help each of them develop a mission statement for their team. Team mission statements should explain the team's purpose for existing and clearly show where the team fits into the larger mission of the overall organization. Figure 7.1 is an example of a team charter for an electronic assembly team for an avionics manufacturing company.

Team Charter
ELECTRONIC ASSEMBLY TEAM

Mission Statement
The mission of the electronic assembly team is to assemble high-quality power supplies for avionics applications as part of the broader manufacturing operations of Avionics Technologies Company.

Team Ground Rules
As we work together to accomplish our mission, all members of the electronic assembly team will abide by the following ground rules:

1. Strive to perform at peak levels on a consistent basis everyday
2. Continually improve our personal performance as well as that of our team
3. Work in mutually supportive ways
4. Exemplify honesty, dependability, punctuality, patience, tolerance, and perseverance
5. Put the goals of the team ahead of personal agendas (selflessness)
6. Resolve conflicts on the basis of what is best for the team
7. Maintain a safety-first attitude
8. Exemplify safety-first work practices

FIGURE 7.1   Sample Team Charter

The mission statement in this example is well written. It is brief and to the point, and it answers the fundamental question a team's mission statement should answer: "What is the purpose of this team?" This team's reason for being is "to assemble high-quality power supplies for avionics applications." Once such a mission statement has been developed for the team charter, the next step is to develop a list of ground rules that specify how team members will work together to accomplish the mission.

Developing the Ground Rules

Whereas the team's mission statement is developed by its supervisor with assistance from human resources personnel, the team's ground rules are developed by the members of each respective team with assistance from the supervisor. A question that always comes up at this point is this: "If employees develop their own team ground rules, how can the supervisor and safety committee make sure that the ones they think are critical—such as safety- and health-related ground rules—are on the final list." This concern is taken care of as part of the process.

While the team members are free to choose their own ground rules, they select them from a list provided by their supervisor and developed with the assistance of the human resources department. The *master list* of ground rules is typically the same for all teams in the organization. In other words, all teams begin with the same list. Consequently, all of the appropriate grounds rules have a chance to make the final list for a given team. This is the first safeguard in the process. The second safeguard built into the process is that the supervisor reserves the right to add ground rules to the final list that were not selected by team members. This approach gives team members a voice in developing their own ground rules, but without management losing control of the process.

The approach I recommend (and use) for developing team ground rules is as follows: (1) the human resources department develops a draft master list for the overall organization, (2) the safety committee submits any safety- and health-related ground rules it wants included in the master list to the human resources department, (3) the draft master list is submitted to higher management for approval, and (4) the approved master list is given to all supervisors to use in guiding their direct reports through a team-level exercise in which the team's ground rules are selected.

Selecting Ground Rules at the Team Level

The supervisor gives each member of his team the master list of ground rules, which typically includes at least all of the items listed above (e.g., honesty, selflessness, mutual supportiveness, initiative, patience . . . safety oriented). It can also contain other items the organization's supervisors, human resources professionals, and higher managers think are appropriate. Each item on the list is presented by the supervisor and discussed with team members as it relates to team performance. Then the team members are asked to (1) circle all items they think should be included in the team charter and scratch through any they do not think should be included and (2) submit the marked-up list without attribution (no names or other identifiers; this approach will encourage frankness).

The supervisor then compiles the final list after adding any he or she or the organization has determined are mandatory inclusions (e.g., safety). The final list is distributed to all members of the team and discussed a final time. This final list is then converted into sentence form, such as the ground rules shown in Figure 7.1. Supervisors may need the assistance of the human resources department in this step. Further, the safety committee should convert the safety-related item(s) into sentence form for supervisors. The end result will resemble Figure 7.1.

Using the Team Charter

A team charter is one of the most powerful tools a supervisor can have. Once the document has been completed, I recommend that the supervisor distribute copies to all team members and have each team member sign his or her copy. The signed original is placed in the personnel file of each respective team member, and a copy is given to individual team members. Then, as the supervisor monitors the daily interaction of his or her direct reports, he or she can use the team charter as a tool for giving corrective feedback when a team member fails to follow the ground rules as well as for recognizing those who do.

An employee who has signed a team charter that contains safety-related ground rules such as those shown in Figure 7.1 will find it more difficult to justify not working safely or not maintaining a positive attitude toward safety. Further, supervisors will find it easier to give corrective feedback to team members who have signed onto a set of ground rules they helped to develop. Although the team charter is only tangentially the responsibility of the safety committee, it is an important tool for helping develop and maintain a safety-first corporate culture in your organization.

Safety and Team Rewards

One of the most commonly made mistakes for organizations is attempting to implement effective teamwork while maintaining an individual-based reward system. If teams are to function effectively, there must be team-based incentives such as rewards and recognition. Further, if teams are to function safely some of the team-based rewards must be based on safety-related performance. Teams function best when an organization's reward-and-recognition program is tied at least partially to team performance as opposed to being tied solely to individual performance. Further, teams function more safely when recognition and other rewards can be earned for working safely and maintaining an exemplary safety record.

If your organization does not include team performance in its reward-and-recognition program and if team performance does not include safety performance, the safety committee has some work to do. Typically, reward-and-recognition programs are overseen by the human resources department. Consequently, ensuring that your organization's reward-and-recognition system includes team-based safety incentives will require the safety committee to work with the human resources department.

Safety and Team Accountability

There is a management axiom that says: "If you want to improve performance, measure it." Accountability is about keeping score. It is about being held accountable for performance—in this case the team's performance. The most effective teams know what their responsibilities are and how their success will be measured. This is why it is important for the safety committee to ensure that team-based accountability includes safety as a performance criterion. Once again, this is achieved by having the safety committee work with the human resources department.

When performance measures are established for teams in your organization, make sure they include safety-related measures. Team-based accountability measures might include such criteria as the following:

- Lost time due to accidents and injuries
- Lost wages due to accidents and injuries
- Property damage due to accidents and injuries
- Medical costs due to accidents and injuries
- Workers' compensation/insurance costs due to accidents and injuries
- Fire losses due to incidents, accidents, and injuries
- Indirect costs of incidents, accidents, and injuries

Effective teamwork is fundamental to competitiveness. Consequently, ongoing team building is a must in today's competitive environment. Safety-first organizations enhance their team-building efforts by including the component of safety in them. Hence, they receive a double benefit from their investment in teamwork: more effective teams and a safer, healthier workplace.

SAFETY-FIRST CORPORATE PROFILE
Noble Corporation

Noble Corporation is one of the leading providers of contract drilling services for the oil and gas industry worldwide. Noble operates 59 mobile offshore drilling units internationally. The company employs a diverse workforce of more than 3,000 people from 38 different countries Noble Corporation has a strong culture of safety in which people are viewed as the company's most valuable asset, and teamwork is seen as being essential to the company's success.

Noble regards its performance in the key areas of safety, health, and environmental management as a critical component of its overall competitive strategy. Consequently, safety, health, and environmental management are fully integrated throughout all levels of the company down to teams and individual team members. Performance in the key areas of safety, health, and environmental management is monitored and measured just as carefully as financial performance. Building teams and making safety, health, and environmental management part of the process are fundamental to Noble's admirable safety record.

Noble's health, safety, and environment and quality (HSEQ) management system is certified to the ISO 14001 Environmental Management System Standard, and all of the company's safety performance standards are based on accepted industry and international guidelines and standards. Nobel strives to be an industry leader in the areas of safety, health, and environmental management and is committed to the concept of continual improvement. For this reason, Noble Corporation has received numerous safety-related awards, including the prestigious Robert W. Campbell Award presented by the NSC.

Source: http://www.campbellaward.org/index.php/site/noble2004/

COLLEAGUE-TO-COLLEAGUE DISCUSSION CASES

CASE 1: How Can We Make Safety Part of Our Team-Building Program?

"We invest a great deal of time in team building, but there is nothing in our program about safety." This statement was made by Nick Odin, chairman of ABC, Inc.'s, safety committee. The safety committee, with the support of the company's executive management team, is trying to establish a safety-first corporate culture at ABC, Inc. The committee and the company have made some progress, but Odin thinks future progress will be limited until all stakeholders understand that (1) safety is one of the most important responsibilities of team members, (2) safety-first work practices and attitudes are easier to maintain in a teamwork-supportive environment, and (3) there is a safety aspect to all of the skills necessary to be a good team player.

Odin knows that all stakeholders at ABC, Inc., need to understand the relationship between teamwork and safety, but he is not sure how to go about developing that understanding throughout the organization. He is especially concerned that supervisors understand the relationship because they play such a critical role in team building. Discussing this issue is the purpose of today's safety committee meeting.

Discussion Questions

1. Have you or your colleagues ever worked in a situation in which safety was built into the team-building process? Have you or your colleagues ever worked in a situation in which safety was not part of the team-building process? Compare the two situations and explain the advantages and disadvantages you found in each.

2. Assume that you are a member of the safety committee at ABC, Inc. What advice would you give Nick Odin concerning how to go about making safety part of the company's team-building program?

CASE 2: We Need to Have Team Charters for All of Our Teams

In the previous case, Nick Odin and his colleagues discussed how to go about making safety part of ABC's team-building program. All agreed there is much to do before that goal could become reality. However, there was unanimous agreement that a major step forward would be to develop team charters for all of ABC's teams. That was the good news. The bad news was that the safety committee was not sure how to proceed in getting team charters developed and safety built into them.

Discussion Questions

1. Have you or your colleagues ever worked in an organization that used team charters? If so, who developed the charters? Was safety built into the charters?

2. Assume that you are a member of the safety committee at ABC, Inc. What advice would you give Nick Odin concerning how to go about having team charters developed and ensuring that safety is built into them?

Key Terms and Concepts

Before leaving this chapter, make sure you understand the following key terms and concepts as they relate to safety and can accurately explain them to people who are not safety professionals.

Open communication	Resourcefulness
Constructive, positive interaction	Punctuality
Mutually supportive approach to work	Tolerance
Respectful environment	Perseverance
Honesty	Team charter
Selflessness	Team rewards
Initiative	Team accountability
Patience	

Review Questions

Before leaving this chapter, make sure that you can accurately and comprehensively, but succinctly, answer the following questions:

1. Explain how safety fits into the responsibilities of team members.
2. Explain how a teamwork-supportive work environment also promotes safety.
3. List five or more essential team-player skills, and explain the safety aspect of each.
4. Explain what a team charter is and how it can be used to promote safety in the workplace.
5. Explain how team rewards can be used to promote safety in the workplace.
6. Specify the connection between safety and team accountability.

Application Project

It is essential that developing safety-first work practices and attitudes be a normal part of the team-building program in your organization. A major component of the team-building process is the team charter. Many organizations do not have team charters. Some do, but the charters do not contain safety-related ground rules. Which of these categories does your organization fit into? Complete whichever of the following projects applies to your organization:

Project Option 1: No Team Charters in Your Organization

Develop a comprehensive plan for establishing team charters in your organization. Safety must be a major component of the charters. Make sure that you include all of the roadblocks you can anticipate and propose solutions for getting around them.

Project Option 2: Team Charters in Your Organization, but No or Only Weak Safety Elements

Examine the team charters that now exist in your organization. If they have no safety-related elements, develop a comprehensive plan for revising the charters to include a strong safety element. If they have safety-related elements, but they are weak, develop a plan for strengthening them. In either case, make sure that you include all of the roadblocks you can anticipate and propose solutions for getting around them.

Chapter 8

Monitor Safety-First Attitudes and Practices

Major Topics

- Daily monitoring
- Developing the safety-first monitoring checklist
- Preparing supervisors and mentors to monitor
- Monitoring on a daily basis
- Giving corrective feedback
- Periodic monitoring follow-up meetings
- Planning corrective action
- Periodically revising the safety-first monitoring checklist
- The safety-first monitoring checklist and nonsafety concerns
- Using monitoring results to inform the performance appraisal process

There is a management axiom that says: "If you want to know if you are losing weight, you have to step on a scale." Of course, this is just another way of saying that in order to make progress, you have to monitor it. This axiom also applies to organizations that are trying to establish a safety-first corporate culture. If your organization wants to know whether it is making progress in establishing and maintaining a safety-first corporate culture, it will have to monitor the safety-related attitudes and behaviors of its personnel on a daily basis. Further, it will have to use what is learned from monitoring to inform the performance appraisal process.

Daily Monitoring

Most people think of monitoring as observing, and this is an accurate perception for as far as it goes. Monitoring does involve observing, but it also entails correcting and reinforcing—both of which should be done on a just-in-time basis. Correcting or reinforcing on a just-in-time basis means doing so when the attitude or behavior in question is first observed, rather than waiting. One of the goals of daily monitoring is to identify both positive and negative attitudes and practices. Negatives should then be corrected immediately and positives reinforced immediately.

When providing corrective feedback, it is important to do so before employees form a habit of doing things the wrong way and before they make the mistake of interpreting their supervisor's silence as approval. When providing positive reinforcement, it is important to do so in the moment so that employees understand exactly why they are being praised and recognized. Whether providing corrective feedback or positive reinforcement, the shorter the time between employees' actions and their supervisor's response, the greater the effect.

Monitoring on a daily basis is the responsibility of executives, managers, safety professionals, supervisors, and mentors, but especially supervisors and mentors because they have the maximum day-to-day contact with employees. Because it is so important for supervisors and mentors to take the lead in daily monitoring, the safety committee must take the initiative to prepare them to be effective monitors.

Monitoring, as recommended in this chapter, amounts to using the old management concept of management by walking around (MBWA) with safety added to the list of things supervisors and mentors look for while walking around. MBWA sounds like an informal method, and it is. However, its informality should not be interpreted as haphazardness or random snooping. The concept of MBWA, although informal in its application, should be well structured in its observations. To ensure that MBWA is sufficiently structured to serve its intended purpose, I recommend the following steps:

- Develop a safety-first monitoring checklist for supervisors and mentors.
- Use the monitoring checklist to prepare supervisors and mentors to conduct effective observations.
- Observe employee attitudes and behaviors on a daily basis.
- Conduct periodic monitoring follow-up meetings.
- Plan any remedial action that may be necessary.
- Periodically update the monitoring checklist.
- Use the results of daily monitoring to inform the performance appraisal process.

Daily interaction with employees gives supervisors and mentors excellent opportunities to observe their attitudes and practices and to reinforce or correct them as appropriate, including as they relate to workplace safety. However, it should be remembered that supervisors and mentors are not safety professionals. Consequently, they may need help in learning what specifically to look for when monitoring safety-related performance and attitudes. The safety committee can help supervisors and mentors understand what safety-related practices and attitudes to look for by developing a simple tool I call the safety-first monitoring checklist.

Developing the Safety-First Monitoring Checklist

Monitoring the safety-related attitudes and behaviors of employees involves using MBWA, a method that should appear informal to those being observed, but which, in reality, is anything but informal. Even though when applied well MBWA will appear informal to employees, it is actually a well-structured concept that is neither random nor haphazard. When attempting to instill a safety-first corporate culture, daily observation of employee attitudes and practices should be well organized and systematic. An effective way to make this happen is to use a checklist that prompts those making the observations to look for specific safety-first attitudes and practices. This is why I recommend that the safety committee develop a safety-first monitoring checklist that supervisors and mentors can use as a guide for making daily observations.

The safety-first monitoring checklist is a tool that supervisors, mentors, and others can use to help focus their observations when monitoring the attitudes and behaviors of employees. The checklist should be a list of an organization's most important expectations relating to safety. Responsibility for developing the checklist belongs to the safety committee. However, since it will be used primarily by supervisors and mentors, I recommend adding several supervisors and mentors to the safety committee on an ad hoc basis to assist in the development of the checklist.

Figure 8.1 is an example of a safety-first monitoring checklist. This example is probably more comprehensive than the one your safety committee will develop since it covers more areas of safety-related concerns than most organizations have to deal with. Consequently, do not think that all items listed in Figure 8.1 must be in your organization's checklist. On the other hand, feel free to add other items that do not appear in the example

---

SAFETY-FIRST MONITORING CHECKLIST

XYZ Corporation

While monitoring employees on a daily basis, look for examples of both safety-related violations and commendations in the following areas of concern:

Safety-First Work Practices

_____ Ergonomics	_____ Lifting hazards
_____ Machine safeguarding	_____ Vision hazards
_____ Falling/impact hazards	_____ Driving hazards
_____ Temperature hazards	_____ Pressure hazards
_____ Electrical hazards	_____ Fire hazards
_____ Industrial hygiene	_____ Confined spaces
_____ Radiation hazards	_____ Personal protective equipment

Safety-First Attitudes

1. Do employees respond positively to safety-related corrective criticism?
2. Do employees willingly correct any inappropriate work practices pointed out to them?
3. Do employees respond positively to prosafety peer pressure?
4. Do employees take the initiative to correct or point out potentially hazardous conditions?
5. Do employees willingly follow established safety precautions and standards?

Comments:

FIGURE 8.1 Sample Safety-First Monitoring Checklist

in Figure 8.1. This figure is provided only as a prototype that can be used as a guide by safety committees as they develop checklists on the basis of specific circumstances and potential areas of concern in their organizations.

Note that the safety-first monitoring checklist in Figure 8.1 consists of three parts: (1) a list of areas of concern relating to safety-first work practices, (2) a list of questions relating to safety-first attitudes, and (3) a section for making comments. The work practices listed in the first part should be those that are specific to your organization. Your safety committee may choose to add other areas of concern to Part 1 of the checklist or remove any shown in Figure 8.1 that do not apply. The same is true for Part 2 of the checklist. The comments section is to be used by supervisors and mentors for recording any notes that may be necessary to jog their memory as they follow up on their observations.

Developing the safety-first monitoring checklist is the responsibility of the safety committee. However, as mentioned earlier in this chapter, it is a good idea to include several supervisors and mentors in the process, and the more influential these personnel are with their colleagues, the better. You want the input of supervisors and mentors as the checklist is being developed, but you also want the supervisors and mentors who help the safety committee to be advocates with their colleagues for using the checklist. The importance of this strategy cannot be overemphasized.

After supervisors and mentors have used a checklist such as the one in Figure 8.1 for a period of time, the criteria will become mentally ingrained and their application second nature to them. However, while the concept is still new, it is best for supervisors and mentors to have a written copy of the checklist readily available for quick reference and periodic review. I recommend that supervisors and mentors make notes on copies of the checklist as they observe their direct reports at work; hence, the comments section of the checklist.

Preparing Supervisors and Mentors to Monitor

Once the safety-first monitoring checklist has been developed in an organization, the safety committee, with the approval and assistance of higher management, convenes a meeting of all supervisors and mentors in the organization to distribute copies and explain how to use the checklist. This meeting is an important step in the process because, if handled properly, it will ensure that all supervisors and mentors (1) appreciate the importance of the task before them, (2) have an opportunity to voice their concerns and ask questions, and (3) understand how to use the checklist for best effect. As time goes by and new supervisors and mentors are added, this kind of meeting must be repeated. This is done periodically forever. A technique I have found effective is to have experienced supervisors and mentors make the presentation to their colleagues.

All three of these benefits are important, but the second one—allowing supervisors and mentors to voice their concerns and ask their questions—is critical. The safety committee needs buy-in from supervisors and mentors. Giving them opportunities to voice concerns and ask questions is an excellent way to win their buy-in. Further, if participants in this meeting raise issues or concerns the safety committee did not think about during the development of the checklist, there is still time to revise it before supervisors and mentors begin using it.

Monitoring on a Daily Basis

Supervisors and mentors can integrate using the safety-first monitoring checklist into their daily observations in the normal course of doing their jobs. The best approach is to make it part of their daily routine. Employees are just like anyone else in that if they think they are being observed, they go into "on-camera mode." The idea is to monitor the work habits and attitudes of employees for the purpose of reinforcing what is good and correcting what is not.

I recommend recording observations in the comments section of the safety-first monitoring checklist. However, a caveat is in order here. Never record anything in the presence of the employees who are being observed. Doing so can undermine the integrity of the monitoring process. Not only will the employees being observed want to know what you are writing down, their behavior might change in the same way it does when people think they are on camera. Taking a few minutes out of each hour in private to jot down any pertinent notes is a better approach.

Another point is important here. Those making the observations should provide reinforcement or correction as appropriate on a just-in-time basis so as to maximize its value. Supervisors and mentors should jot down their notes later, but praise or correct immediately. Public praise should be extended the moment an employee is observed satisfying any criterion on the safety-first monitoring checklist. Correspondingly, private correction should be given any time an employee's attitude or behavior is at odds with any of the criteria on the checklist. Notes taken by observers are for use when following up on daily observations and later when conducting performance appraisals.

Giving Corrective Feedback

Providing corrective feedback is a critical responsibility of supervisors as part of the monitoring process. Corrective feedback is another term for *constructive criticism.* Consequently, it is important that it be given and received in a constructive way. Constructive criticism that is poorly delivered may be received by employees as just plain criticism. For this reason, the safety committee should ensure that communicating corrective feedback is covered in the training your organization provides for supervisors.

Since supervisors must give corrective feedback to employees on more performance-related issues than just safety, this essential supervision skill will probably be covered in the generic supervision training provided by your organization. However, the safety committee should make sure that it is. The following guidelines will help supervisors ensure that corrective feedback is received in a positive manner.

Begin with Something Good

A complaint frequently made by employees about supervisors can be paraphrased as follows: *You always tell me what I'm doing wrong, but you never tell me what I'm doing right.* Whether or not this kind of complaint is valid depends on the individual supervisor. However, it does point out the underlying problem with giving corrective feedback.

People tend to view constructive criticism as just plain criticism if all they ever hear about is the need to improve their performance.

To help ensure that constructive criticism is accepted in a positive manner, supervisors should begin with something good. Point out something the employee in question does well and compliment him or her on it. Use the compliment as the lead-in to the corrective part of the message. For example, a supervisor might say the following to a team member: "John, you are doing an excellent job in holding down the amount of scrap you produce. Now, let's talk about how you can do equally well in following our safety procedures."

The bottom-line message conveyed in this statement is that John needs to follow the organization's established safety rules. However, before making this point, the supervisor let John know that he is doing a good job in reducing scrap. This is important because (1) it lets John know that the supervisor is noticing his good work, and (2) it builds John up so that he is better able to accept the message that he needs to do a better job of following the safety rules.

Have the Facts

When they hear the message "You need to improve," some people will respond with denial. This is especially true of members of the entitlement generation, the 18- to 30-year age group that grew up more sheltered from accountability than previous generations. According to Jean M. Twenge, author of *Generation Me*—the groundbreaking book on the entitlement generation—one of the characteristics of people in this age cohort is an aversion to criticism, even when it is constructive and well intended.[1]

"Me-geners" grew up in a time when parents routinely intervened to protect them from the consequences of their actions (accountability), teachers were often more concerned with developing self-esteem than teaching responsibility, everybody made the team in sports, everybody on the team got to play, and so on. In other words, for members of the entitlement generation, receiving corrective feedback might be an alien concept.

For people unaccustomed to it—such as me-geners—the first reaction to corrective feedback is often denial. Consequently, supervisors need to understand that they must be prepared with the facts. This is another reason why the safety-first monitoring checklist is such a valuable tool. As supervisors jot down notes and dates on checklists, they are creating a factual record of their observations and corresponding interactions with employees. Then, when employees react to corrective feedback with denial, the supervisors have the facts to make their point.

Use Good Judgment

Part of giving corrective feedback involves setting goals or targets for measuring improvement. A supervisor needs to understand that good judgment is important when establishing improvement targets. This is where the supervisor's experience comes into play in a valuable way. For example, if the employee in question is engaging in work practices that are dangerous—not wearing a hard hat in designated areas—corrective action needs to be 100 percent and immediate. On the other hand, if a new employee is turning out just 5 parts per day on a milling machine and should be making 10, the supervisor is unrealistic to expect 100 percent improvement in just one day. Improvement in any performance area is important, especially in the area of workplace safety, but knowing how much improvement to expect and how soon requires good judgment.

Periodic Monitoring Follow-Up Meetings

The safety committee should periodically convene monitoring follow-up meetings with supervisors and mentors to discuss what they are seeing as they observe employees at work. The goal is to identify trends or recurring problems. During periodic monitoring follow-up meetings, the chair of the safety committee leads supervisors and mentors through a criterion-by-criterion discussion of their observations covering a specified period of time. Where are the most problems? Are we making progress? Where do we need to focus our remedial efforts? Questions such as these are asked by the individual leading the meeting, and the answers are discussed and debated by participating supervisors, mentors, and safety committee members.

A running tally of problems, issues, and concerns is made by whomever the chair of the safety committee appoints to take notes. A simple but effective way to accomplish this task is to use flip charts with pages that can be removed and affixed to the wall of the conference room so that group feedback is visible throughout the meeting. The idea is to identify problems and trends that the majority of observers found, rather than isolated issues, good or bad. The list of the most widely and frequently observed problems, issues, and concerns recorded during such meetings will become the basis for the plan that is developed in the next step.

Planning Corrective Action

After each monitoring follow-up meeting, supervisors, mentors, and the safety committee should reconvene for the purpose of planning corrective action for trends that have been identified or problems that are recurring. Problems that are frequently seen by the majority of observers as well as any negative trends observed become the focus of the planning session. For example, assume that employees are frequently observed improperly lifting heavy bulky objects. If this is the case, training in proper lifting techniques might be called for.

SAFETY-FIRST FACT
To Improve Safety Performance, Measure It

According to publishing giant Pearson, measuring safety performance is an essential part of continually improving it. To this end, Pearson has adopted a framework for consistently collecting and displaying safety-related data throughout its many facilities worldwide. This framework consists of the following elements: (1) fatalities, (2) major injuries and illnesses, (3) all other injuries and illnesses, (4) work days lost, and (5) enforcement notices or prosecutions. Having consistent data from all of its various facilities allows Pearson to establish benchmarks for measuring performance and determining when real improvements have been made.

Source: Retrieved from http://www.pearson.com/index.cfm?pageid=148 on January 31, 2008.

If all employees have completed the required training in proper lifting techniques but are ignoring what they were taught, perhaps there is an attitudinal problem that must be dealt with. As an alternative to training, or even in addition to it, mentors and supervisors might be tasked with focusing more attention on attitudes toward proper lifting as they interact with their protégés.

The plan developed is put into action immediately. Daily monitoring and periodic monitoring follow-up meetings continue in the normal manner. In fact, they become the organization's mechanism for determining how effective the plan developed in this step has been in solving the problems identified. Periodically new corrective plans are developed and implemented, and the cycle repeats itself on a continual basis.

Periodically Revising the Safety-First Monitoring Checklist

The organization's safety-first expectations provide the basis for developing the safety-first monitoring checklist, as was explained earlier. This does not change. However, because of changes that occur in the nature of work in your organization over time—for example, manual processes that become automated, new equipment, or an added process to accommodate a new product line—it might be necessary to periodically revise the safety-first monitoring checklist to accommodate these changes.

Revising the safety-first monitoring checklist is the responsibility of the safety committee. However, the process should include input solicited from supervisors and mentors. In addition to changes in the nature of work that occur over time, experience in using the checklist will occasionally suggest revisions that will make it a more effective tool. Regardless of why the checklist needs to be revised, it is important to keep it up to date.

The Safety-First Monitoring Checklist and Nonsafety Concerns

The job of supervisors is to secure consistent peak performance from their direct reports and continually improve that performance. This means that supervisors lead, guide, and coach their team members in ways that will help them perform at their best in terms of all three elements of superior value: quality, cost, and service. One of the ways supervisors can help their team members achieve consistent peak performance and continual improvement of quality, cost, and service is to provide a safe and healthy work environment. This concept was established earlier in this book as one of the reasons for developing a safety-first corporate culture.

As I work with organizations that are trying to establish a safety-first corporate culture, supervisors always raise a specific issue. Their comments about this issue can be summarized and paraphrased as follows: *I can see the value of looking for safety-related attitudes and practices as I use MBWA to monitor my team members. But safety-related attitudes and practices are only part of what I need to monitor. I also need to look for quality, cost, and service-related attitudes and practices.* This is a legitimate concern and one safety professionals will probably have to deal with.

When this issue is raised by supervisors, I tell them that the reason for developing and using the safety-first monitoring checklist is that they may not be as accustomed to looking for safety-related concerns as they are to looking for other performance-related concerns such as quality, cost, and service; hence, the need for the checklist. Once the items on the safety-first monitoring checklist become as deeply ingrained in their minds as the other performance indicators they look for, they will no longer have to use the checklist or they will need to simply review it periodically to refresh their memory.

I follow this explanation up with an important caveat that goes like this. If you are a new supervisor for whom quality, cost, and service expectations are not yet second nature, those expectations may be added to the safety-first monitoring checklist to create a more comprehensive, integrated tool—a tool that can be used for monitoring all aspects of employee performance. In fact, for new supervisors this is probably a good idea.

Remember that safety is not an isolated, stand-alone concern in the workplace. Rather, it is a foundational concern that can affect all other areas of performance as they relate to your organization's competitiveness. Consequently, there is nothing wrong with transforming the safety-first monitoring checklist into a broader, more comprehensive tool for use by supervisors in monitoring all aspects of employee performance. Just make sure that if you take the broader, more comprehensive approach, safety-related criteria do not get lost or left off.

An example of when it might be necessary or wise to broaden the scope of the safety-first monitoring checklist can be found in Coachmen Industries, Inc. Coachmen is a leading manufacturer of recreational vehicles, modular homes, and associated products, including motor homes, travel trailers, fifth wheel trailers, folding camper trailers, fiberglass and thermoform plastic products, and modular commercial buildings. The company's Housing and Building Systems Group is the leading producer of modular homes in the United States.

Coachmen Industries operates in highly competitive markets. Consequently, in order to succeed, the company must maintain a corporate culture that emphasizes both performance and safety. To this end, in addition to core values relating to safety, the company has the following quality- and service-related core values:

- Providing customers with quality products and service
- Treating dealers and suppliers as partners
- Pursuing customer satisfaction unrelentingly

In monitoring team members on a daily basis, Coachmen's supervisors are responsible for observing safety-related attitudes and practices, but they are also responsible for observing the quality- and service-related performance criteria. Consequently, it would be entirely appropriate for supervisors—particularly new supervisors—at Coachmen Industries, or any other company for that matter, to include criteria relating to nonsafety performance concerns among the indicators they look for when monitoring employees.

Using Monitoring Results to Inform the Performance Appraisal Process

As was stated earlier in this chapter, to know if you are losing weight, you have to step on a scale. To organizations trying to instill a safety-first corporate culture, this maxim means that employee attitudes and practices toward safety must be monitored daily. But just monitoring—although critical—is not enough. The appropriate attitudes and practices must be reinforced and the inappropriate corrected. This is done in real time as part of the monitoring process. However, high-performing organizations with safety-first corporate cultures derive a double benefit from the process by using the results of monitoring to inform the performance appraisal process.

Including safety-related criteria in your organization's performance appraisal instrument was explained in Chapter 2. Deciding how to rate employees on these criteria is the issue in this chapter. If supervisors in your organization have used the safety-first monitoring checklist for any period of time prior to conducting periodic performance appraisals, they should know exactly how to rate their direct reports on safety-related criteria. This is because by monitoring safety-related attitudes and practices on a regular basis and by reinforcing or correcting them as appropriate, supervisors have all of the information they need to assign ratings on performance appraisal instruments.

For example, if a supervisor finds it necessary to correct a given employee several times for the same inappropriate attitude or practice, the employee's performance appraisal should reflect this fact. On the other hand, an employee who has been commended more than once for exemplifying safety-first attitudes and practices should receive correspondingly positive ratings. In both cases, copies of the safety-first monitoring checklist can serve as both a reminder and documentation.

A more common—and more desirable—situation is the employee who has been corrected and has responded by improving. The performance appraisal should reflect the fact that corrective action was necessary and was taken and that satisfactory improvement has been made. This is what is meant by using monitoring results to inform the perform performance appraisal process.

SAFETY-FIRST CORPORATE PROFILE
Kellogg's

Kellogg's is one of the most widely recognized corporate names in the world. Beginning with just 44 employees in Battle Creek, Michigan, in 1906, Kellogg's now employs more than 30,000 people in 18 countries. The company's products are sold in more than 180 countries. With earnings of more than $12 billion annually, Kellogg's is the world's leading producer of cereal and one of the world's leading producers of convenience foods. Such a large and diverse global organization must be concerned about the safety and health of its workforce, and Kellogg's is.

The corporate values of the company—known as "K Values" and which include workplace safety and health—are at the heart of its commitment to corporate social responsibility, a fact that led to Kellogg's being selected as one of the world's most ethical companies. Its corporate social responsibility committee, established in 1979, is responsible for the practical application of the organization's commitment to safety, health, and environmental management. Monitoring safety-related attitudes and practices is an important part of the company's overall approach to ensuring a safe and healthy work environment for its personnel.

Supervisors at Kellogg's plants worldwide are trained to monitor the safety-related work practices of their direct reports and to take immediate corrective action if their observations reveal the need. At Kellogg's, safety is a high priority at all levels of the organization, and daily monitoring is just one way the company carries out its commitment.

Source: http://www.kelloggcompany.com

COLLEAGUE-TO-COLLEAGUE DISCUSSION CASES

CASE 1: How Are We Going to Get Our Supervisors Involved?

As soon as the meeting of the safety committee for ABC Construction came to order, one of the members asked, "How are we going to get our supervisors involved in monitoring safety practices and attitudes?" The person asking the question was Casey Aland, one of ABC Construction's supervisors. Aland represented all of the company's supervisors on the safety committee. His question prompted a long discussion of the various reasons why the company needed its supervisors to add safety to the other employee practices they monitor on a daily basis.

Aland listened while the other members justified adding this responsibility to what is expected of the company's supervisors. Finally, he held up his hand to silence the discussion and said, "I did not ask why we need to get supervisors involved. I understand the *why* part of the discussion. I asked *how* we were going to get supervisors involved." Aland went on to explain that the company's supervisors were not safety experts and might not know what to look for when monitoring. He also expressed concern that ABC Construction's mandatory training for supervisors did not go into much detail concerning safety and health issues.

Discussion Questions

1. Have you or your colleagues ever worked in a situation in which supervisors played little or no role in monitoring safety-related practices and attitudes on a daily basis? If so, describe the situation and any problems associated with it.

2. Put yourself on ABC Construction's safety committee. How should the committee go about getting the company's supervisors involved in effectively monitoring safety-related work practices and attitudes?

CASE 2: We Need to Make This Part of the Performance Appraisal Process

ABC Construction's safety committee from Case 1 above was able to develop a comprehensive plan for getting supervisors actively involved in monitoring safety-related work practices on a daily basis. The plan had been implemented and, after a few initial glitches were worked out, appeared to be working well. In fact, ABC Construction's safety committee was now holding its first monitoring follow-up meeting.

During this meeting, one of the members of the safety committee, commenting on that month's monitoring results, said, "We need to make this part of the performance appraisal process." After some discussion, the safety committee decided that this was a good idea, at least conceptually. "But," said one member, "we need to discuss how to go about doing this."

Discussion Questions

1. Have you or your colleagues ever worked in a situation in which the results of safety monitoring were not included in the performance appraisal process? If so, describe the situation.

2. Put yourself on ABC Construction's safety committee. How should the committee go about getting the results of daily monitoring included in the company's performance appraisal process?

Key Terms and Concepts

Before leaving this chapter, make sure you understand the following key terms and concepts and can accurately explain them to people who are not safety professionals.

Daily monitoring	Begin with something good
Management by walking around (MBWA)	Have the facts
Safety-first monitoring checklist	Use good judgment
Preparing supervisors and mentors to monitor	Monitoring follow-up meetings
Monitoring on a daily basis	Planning corrective action
Providing corrective feedback	Using monitoring results to inform the performance appraisal process
Constructive criticism	

Review Questions

Before leaving this chapter, make sure you can accurately and comprehensively, but succinctly, answer the following review questions:

1. Explain the concept of daily monitoring as it relates to maintaining a safety-first corporate culture.

2. Explain how an organization should go about developing a safety-first monitoring checklist.

3. Describe how an organization might go about preparing its supervisors and mentors to monitor safety-first attitudes and practices.

4. Explain how supervisors and mentors can give corrective feedback so that it is well received.

5. Describe what happens in a monitoring follow-up meeting.

6. State why it is important to periodically update the safety-first monitoring checklist and how an organization should go about it.

7. Discuss whether it appropriate for the safety-first monitoring checklist to contain nonsafety concerns that should also be monitored.

8. Explain how an organization can use the results of daily monitoring to inform the performance appraisal process.

Application Project

In safety-first organizations, supervisors play a key role in making sure the workplace is a safe and healthy environment. Supervisors can be the safety committee's best allies or worst enemies when it comes to establishing and maintaining a safety-first corporate culture. Consequently, it is important for the safety committee to effectively enlist supervisors into the "cause," something that can be easier to say than do. To this end, your safety committee is tasked with the following:

1. Develop a safety-first monitoring checklist for your organization
2. Develop a comprehensive implementation plan that contains the following elements:
   a. Preparing supervisors and mentors to monitor
   b. Including a section on giving corrective feedback in training for supervisors
   c. Using monitoring results to inform the performance appraisal process

Endnote

1. Jean M. Twenge, *Generation Me* (New York: Free Press, 2006), 150–56.

Chapter 9

Reinforce Safety-First Attitudes and Practices with Rewards and Recognition

Major Topics

- Rationale for reinforcing safety-first attitudes and practices
- Methods for reinforcing safety-first attitudes and practices
- Rationale for developing an integrated reinforcement system
- Adopting the menu approach for recognizing personnel

There is a maxim that says: "You usually get what you expect." This maxim is the basis for Chapter 2, "Expect Appropriate Safety-First Attitudes and Practices." As was explained in Chapter 2, it is important to let people in organizations know what is expected of them. Employees who engage in unsafe work practices are not solely to blame if your organization has done nothing to show them that safe work practices are expected. However, setting high expectations represents just one side of the human performance coin. The other side involves reinforcing behaviors that accord with the expectations. This chapter explains how organizations can build safety into their formal recognition-and-reward program so that safety-first attitudes and practices are properly reinforced.

Rationale for Reinforcing Safety-First Attitudes and Practices

People have an inherent need to know how they are doing and to feel they are appreciated. These two inherent human needs, taken together, form the rationale for reinforcing safety-first attitudes and practices. This is why it is so important to recognize and reward (reinforce) personnel who exemplify an organization's expected attitudes and practices. When people in positions of authority in the organization recognize and reward positive attitudes and behaviors, they are telling their personnel, "You are doing a good job of exemplifying our expectations," and "you are appreciated." This is important because coupling expectations with reinforcement is the best-known way to influence human behavior in a positive manner.

A story from my college days illustrates how powerful reinforcement in the form of recognition and rewards can be. My classmates and I were taking a graduate class on evaluation methods. The professor had lectured in the two previous classes about the importance of reinforcement as part of the evaluation process. To illustrate his point, he asked for volunteers to prepare a group presentation to give to the class. The three of us who volunteered to make the presentation were given the following expectations: (1) be positive and upbeat, and (2) spread your attention over the entire class. Without telling the presenters what he was doing, the professor told the rest of the class that everyone on the right side of the class (the speaker's right) should give off nonverbal signs of ignoring the speaker. He told the students on the left side of the class to smile, nod, and give off other signs of not just paying attention, but agreeing with the speaker.

As I began my part of the presentation, I noticed immediately that everyone in the audience on my right seemed to be half asleep—not even one of them seemed interested in what I had to say. I remember thinking, "Wow, could I be doing that badly?" However, the students on my left were paying careful attention to my every word, seemed openly receptive to my message, and appeared to agree with everything I was saying. Remembering the professor's expectations, I tried to stay positive and upbeat when looking out at the right side of the class, but the complete lack of interest I saw there made it increasingly difficult. Without realizing it, the longer the presentation went on, the more time I spent in front of the left side of the class and the more attention I gave to the students who were giving me positive reinforcement. My two fellow presenters had the same experience.

When the presentation finally came to a merciful end, the professor asked the three of us to describe for the class how we felt about the reinforcement received from the audience. After listening to our impressions, he asked if we realized that within five minutes of beginning to speak, each of us in succession had become firmly rooted on the left side of the class, ignoring the students on the right side. He then explained that the point of the exercise was to demonstrate in a memorable way that people respond to reinforcement. They are more likely to do what is expected if what is expected is reinforced. He explained that we were expected to be positive and upbeat and to spread our attention around evenly over the entire class. What actually happened was that we did what was expected only when it was reinforced in a positive way. The smiles, nods, and attention of the left side of the room helped us be positive and upbeat and to give that side of the room our full attention. However, the lack of reinforcement from the right side of the room caused us to not just ignore that side, but to avoid it altogether. The professor's demonstration made a powerful impression on the class, and I have never forgotten it. His message is just as applicable in organizations trying to establish a safety-first corporate culture as it was in that college classroom many years ago. Reinforce what you expect of people so they know how they are doing and that they are appreciated.

Methods for Reinforcing Safety-First Attitudes and Practices

There are two methods for reinforcing human performance that are especially effective in the workplace: rewards and recognition. In the current context, rewards take the form of monetary reinforcement (e.g., salary increases, bonuses, cash incentives). Recognition involves saying "good job" in a formal way, using nonmonetary reinforcement methods. There is an important caveat that needs to be understood here.

Nonmonetary reinforcement does not necessarily mean that the method chosen does not have a price tag. In fact, there is often a monetary cost to the organization for nonmonetary reinforcement methods. The difference between monetary and nonmonetary reinforcement is that with the former, dollars are actually exchanged. For example, the employee or team in question receives a salary increase, bonus, or cash incentive for doing an exemplary job of meeting and exceeding expectations. With the latter, the employee or team receives some form of reward or recognition such as a gift card, tickets to a movie, or passes to an amusement park. These items all have an associated cost that the organization must pay, but the employee or team in question receives a gift card, tickets, or passes rather than actual dollars.

Most organizations have some form of reward-and-recognition system for reinforcing excellent performance. The key, in the current context, is to make sure that safety is a key criterion in selecting the employees to be rewarded or recognized. To do this, your organization's safety committee will once again have to take the lead, but work closely with the human resources department and higher management.

Rewarding Individual and Team Performance Relating to Safety

The safety committee's first task is to review how the organization currently goes about choosing the individuals and teams that are to be singled out for performance-based rewards. For example, an individual's performance evaluations over time are typically considered when deciding whether he or she warrants a raise or promotion. Consequently, it is important to ensure—as was explained in detail in Chapter 8—that your organization's performance appraisal forms contain safety-related criteria. If performance bonuses or cash incentives are provided, there should be clearly identifiable criteria. The safety committee's task is to make sure that these criteria include safety-related practices and attitudes.

An employee who performs well on most criteria, but poorly on safety, should not be promoted or given a raise, bonus, or cash incentive. Correspondingly, outstanding performance in safety-related practices and attitudes should be rewarded. The most effective way to make sure this happens is to include safety as a fundamental criterion when making decisions about performance-related rewards. If safety is not currently included in your organization's criteria for performance rewards, the safety committee should work with the human resources department and higher management to revise the criteria appropriately.

Recognizing Individual and Team Performance Relating to Safety

While ensuring that safety-related attitudes and practices are part of the criteria for giving performance-based rewards, the safety committee should also ensure that they are considered when singling out individuals for recognition. For example, if your organization recognizes individuals and teams with awards such as "employee of the month" or "team of the year," part of the criteria considered when selecting the individuals and teams to be recognized should be safety-related performance.

The safety committee's responsibility in this is to work with your organization's human resources department and higher management to make sure that (1) safety-related attitudes and practices are part of the selection criteria, and (2) no person or team with a poor safety record will ever be singled out for recognition. Later in the chapter, I explain what I call the "menu approach" for giving recognition awards. The important point here is that criteria relating to safety attitudes and practices must be fully integrated into the selection process.

Safety-first companies use rewards and recognition as part of their overall strategy for maintaining a safe and healthy workplace. When organizations are rewarded or recognized for their outstanding safety record, as safety-first organizations often are, one is safe in assuming that they have a reward-and-recognition system that includes safety-related criteria. I have never dealt with an organization with an award-winning safety record that did not build safety and health into its employee reward-and-recognition program.

An excellent example of such an organization is Anheuser-Busch Company. This global beverage company has been recognized many times by a variety of different organizations and agencies. What follows are just a few examples of the safety and health awards won by Anheuser-Busch:[1]

- Perfect Record Award from the NSC for outstanding safety performance when 40 of its facilities went a full year with no disabling injuries.

- Green Cross Award from the NSC when 67 of its facilities had less than half of the national average for their industry group in terms of lost time due to injuries.

- Chairman's Merit Award from the Safety Center, Inc., to the Fairfield Brewery facility in Fairfield, California, for commitment to safety in the workplace.

- Gold Medal Safety Award from the Professional Association for Agriculture and Forestry in Franconia and Upper Bavaria for its Huell Hop Farm in Huell, Germany.

- Safe Production Model Employer Award from the Wuhan Safe Production Committee of the Wuhan Municipal Government to the Wuhan Brewery in Wuhan, China.

The point here is not that Anheuser-Busch wins so many safety awards—as commendable as that fact is. Rather, the point is that the company wins awards and recognition by beginning at the individual and team levels and making safety and health part of the organization's performance criteria as well as its criteria for giving rewards and recognition to both individuals and teams. To establish and maintain a safety-first corporate culture, your organization will need to do the same.

Rationale for Developing an Integrated Reinforcement System

I sometimes find myself advising organizations that are so determined to establish a safe and healthy work environment that they segregate safety and approach it as a stand-alone concept, rather than fully integrating it into their policies, processes, and practices. This often happens as the result of a major safety disaster or a large fine that has been assessed by OSHA. Although I commend such organizations for their determination to emphasize the importance of workplace safety, the stand-alone approach is a mistake in my opinion. In fact, you may recall from the Introduction that the philosophical foundation of this book is that safety and health should be fully integrated concepts.

In practical terms, appendages are the easiest to cut off when organizations face difficult financial times. Stand-alone safety and health programs or stand-alone aspects of such programs can easily become inviting targets for budget cutters when financial hard times force organizations to tighten their belts. But safety and health policies, processes, and procedures that are fully integrated cannot be so easily cut off and cast aside. Further, safety and health will never rise above step-child status as long as it is treated as a stand-alone entity in your organization. Finally, as was shown in the Introduction, safety and health policies, processes, and procedures are more effective—they work better—when they are fully integrated as part of an organization's normal operations.

Consequently, I recommend against giving separate, stand-alone recognition awards for safety, unless, of course, these awards can somehow be incorporated as fully integrated aspects of an organization's overall performance reinforcement program. Rather, I recommend that organizations incorporate safety-related criteria into the selection process for recognition awards and into the decision-making process for performance rewards such as raises, promotions, bonuses, and incentives. When this approach is followed, safety becomes a normal and integral part of an organization's operations, rather than a stand-alone appendage that can be easily lopped off whenever financial belt tightening is needed.

As should be abundantly clear by now, safety professionals and safety committees have to work closely with supervisors, human resource personnel, and higher management to establish and maintain a safety-first corporate culture. My experience has shown that it is less difficult to gain the commitment of supervisors, human resource personnel, and higher management to policies, processes, and procedures that are fully integrated within organizations. Stand-alone entities in organizations always seem to retain an adjunct feel about them. Further, safety will never become a fully integrated aspect of your organization's culture until safety policies, processes, and procedures are fully integrated.

How this fact comes into play in an organization can be illustrated by an example from my experience. I once consulted with a company that gave separate safety awards outside of its normal reward-and-recognition program. Unfortunately, the awards never seemed to gain the recipients any status or esteem among their colleagues. In fact, employees at this company were more concerned about receiving the various performance awards that made up the company's organization-wide reward-and-recognition program. There was even a separate award ceremony for the safety awards. Consequently, recipients already felt like outsiders. They won the esteem of the safety committee, but no one else.

Further, occasionally an individual who was not an especially high performer on other aspects of his or her job would receive a safety award. When this happened, the value of the safety awards was diminished even further. Once, during a safety awards ceremony, I overheard an employee comment, "If he can win an award, anybody can." This employee was talking about a teammate who had received a stand-alone safety award. In a safety-first organization, safety is one performance criterion considered when selecting the recipients of rewards and recognition. Of course, it is a critical criterion and one that I think should be "weighted" more heavily than others. But it is not a stand-alone criterion or award.

There is also another reason I recommend that rewards and recognition for safety be fully integrated into your organization's policies, processes, and procedures. Your organization's willingness to cooperate in the integration is an indicator of its real commitment. It is always easier to just add another organizational appendage than to go through all of the discussion, debate, retraining, and revisions necessary to integrate policies, processes, and procedures. Integration requires the cooperation of supervisors, human resource personnel, and higher management. It requires revising job descriptions, team charters, performance appraisal forms, policy manuals, procedures manuals, and other forms of documentation. It requires patience, persistence, perseverance, collaboration, cooperation, and constant attention. In short, integration is the hard way. However, in this case, as is often true, the hard way is the right way.

Adopting the Menu Approach for Recognizing Personnel

This section explains how your organization can adopt the menu approach for recognizing personnel. This section assumes that your organization had adopted an integrated approach to rewards and recognition as recommended in the previous section. However, before explaining the menu approach, a few words about recognition as a tool for reinforcing human performance is in order. All members of the safety committee need to understand the use of recognition as a tool for reinforcing human performance because they might have to explain it to human resource personnel and higher management in your organization.

The first thing to understand is that recognition is typically a better motivator and reinforcer than financial rewards such as raises, bonuses, and incentive dollars. This may be difficult for you to believe, but in the long run it is true. Financial rewards are more effective for *maintaining* personnel than for *motivating* them to perform better. In other words, financial rewards are more effective as retention tools than as motivation and reinforcement tools. The reason for this is that people who receive a raise, bonus, or incentives dollars soon become accustomed to having the extra money. Consequently, before long the new dollars just become an expected part of the employees' compensation package, at least in their minds. The way I explain this phenomenon is that employees who receive financial rewards are typically motivated by them for about three weeks. After that they have usually spent the money and are right back where they were before receiving it, just at a higher level.

Recognition, on the other hand, is not expended. In fact, it tends to grow more valuable over time. The plaque on the wall, certificate in the personnel file, or other visible indicator of the recognition received—and the esteem that goes with it—stays with employees forever and cannot be taken from them. For example, on the wall of my study hangs a framed newspaper article about an award I once received. That award came with a $1,500 cash bonus, nice plaque, and a newspaper story. I don't even remember what I did with the money, although I'm certain I spent it fairly soon after receiving it. But that plaque and the accompanying framed newspaper story are still with me and always will be. People see them in my study and ask about them. People in my field immediately recognize the award and what it represents. If I had my choice between the plaque and the $1,500 cash bonus, I would take the plaque. I relate this story to make the point that one of the most effective methods for reinforcing safety-first attitudes and behaviors as part of a broader reward-and-recognition program is to recognize individuals and teams. If they secure raises, bonuses, or incentive dollars as a result of their outstanding performance, good. But do not make the mistake of relying solely on financial rewards to reinforce your organization's performance expectations. Systematic recognition is essential to effective reinforcement. To understand just how essential, consider the following findings from a study conducted by the Minnesota Department of Natural Resources that found that recognition contributes to both motivation and job satisfaction[2]:

- Most respondents said they highly valued day-to-day recognition from their supervisor and peers.
- A majority of respondents (68 percent) said it was important to them to know that their work was appreciated.
- A majority of respondents (63 percent) said that most employees would like more recognition for their work.

- A majority of respondents (67 percent) said that most people need to be appreciated for their work.
- Only a small minority of respondents (8 percent) said that employees should not expect to be praised for their work.

This information is provided as background to ensure that all members of the safety committee understand how important it is to make systematic recognition of personnel one of the key methods used for reinforcing safety-first attitudes and practices in your organization. However, I have found that most of the traditional recognition methods are limited in their effectiveness by one glaring weakness: With people, one size never fits all. I developed the menu approach for recognizing personnel to counteract this weakness.

Overview of the Menu Approach

The menu approach will work in conjunction with almost any established recognition program your organization might already use. For example, if your organization recognizes outstanding performance by selecting employees of the month, quarter, or year, the employees in question typically receive some type of predetermined reward, such as a parking space, plaque, framed certificate, mug, tee shirt, or hat. With the menu approach, rather than awarding the same predetermined amenities to all employees who are recognized, the organization gives the winners a menu of rewards from which they may choose their prize. If a winner wants a parking space, he or she may select it from the menu. However, if a winner prefers some other amenity, there are other options.

For example, assume an employee named Mark Stone has performed especially well for ABC, Inc., in the areas of safety, quality, customer service, and waste reduction during the past quarter. The management team at ABC, Inc., decides to name Stone its "Employee of the Quarter." Rather than give Stone a parking space or a plaque, the organization gives him a menu from which he may select one of approximately 25 prizes—rewards that include a parking space, plaque, company shirt, gift cards, and movie tickets. The beauty of the menu approach is that nobody knows better than Stone, himself, what kind of amenity will be the most effective reinforcer and motivator.

Most organizations I have worked with use the traditional one-size-fits-all approach when recognizing their personnel. The problem with the traditional approach is that regardless of the individual differences among them, all personnel get the same reward (plaque, certificate, parking space, etc.). The fundamental weakness of this one-size-fits-all approach is that what motivates one person to perform better in terms of safety or any other performance indicator may not necessarily motivate another. Further, the standard amenities that accompany recognition tend to lose their effect over time—they go stale.

An especially high-performing employee once showed me a wall full of plaques and certificates he had earned to commemorate the recognition he had received at different times. When I commented that he "really had a lot of plaques," this high performer said, "I know. After a while they all just seem to blend together. My company needs to find something different to give the employees it wants to recognize. Don't get me wrong. I appreciate the recognition, but how many plaques does one person need?" He had a point.

This individual's point was reinforced for me many years ago when, while traveling to a consulting job, I pulled off Interstate 10 at a fast-food restaurant between Houston and

San Antonio, Texas. At the time, this section of interstate highway was especially desolate. You could drive for miles and see nothing but dust and tumbleweeds. In fact, the fast-food restaurant where I stopped was like an oasis in the middle of the desert, and it had one of the largest parking lots I have ever seen. Drivers of 18-wheelers used the parking lot as a convenient place to stop and catch up on their sleep because the parking lot was so expansive it swallowed up these big-rigs to the point that one hardly noticed their presence.

I parked in a space near the front door. As I got out of my car, I noticed the parking space next to mine was reserved for the restaurant's "Employee of the Month." There was a bicycle parked in the space. While eating my lunch, I noticed a young man cleaning tables who had a blue ribbon attached to his name tag that read "Employee of the Month." I engaged this young man in conversation and asked if the reserved parking space near the front door was his. He acknowledged that it was and then laughed as he said, "Some motivator, huh?" As a business major at a nearby college, the young man understood that the restaurant was trying to recognize him for outstanding performance, that it wanted to reinforce excellence and motivate him to keep up the good work as an example to his teammates. But he was not impressed with his organization's choice of reward. His final comment to me was revealing: "If they really wanted to motivate me, they should have given me a more flexible schedule so it wouldn't be so hard to enroll in the classes I need. I don't need a parking space, but I do need a more flexible work schedule."

This comment summarized succinctly, but powerfully, the weakness of most recognition systems in organizations, a weakness I was already aware of and had thought some about. The young man appreciated the thought behind the recognition given him, but he hardly needed a reserved space for his bicycle, when the restaurant had a parking lot the size of Nebraska. For him, a reserved parking space was not a motivator. However, a more flexible schedule would have been. This example helped me firm up a growing conviction that, when it comes to the rewards given to employees being recognized for outstanding performance—rewards you hope will reinforce that performance—one size does not fit all. This fact is the principal reason I recommend the menu approach for recognition programs.

Developing the Recognition Menu

For the menu approach to be effective, it is necessary to involve employees in both its initial development and periodic updating. The reason for involving employees in developing and periodically updating the recognition menu goes beyond just buy-in. Buy-in to the recognition process on the part of employees is, of course, important. But even more important than buy-in is the need to make sure that the items on the menu will do what your organization needs them to do: reinforce the expected attitudes and practices in those who are being recognized and motivate other employees to follow their example. The best way to know that menu items will truly appeal to employees is to let employees decide which items the menu will contain. I have found the following process to be effective for developing recognition menus:

Step 1: *Assign the Task of Developing the Recognition Menu*
If the safety committee is broadly representative of your organization's various functional units and includes a member from the human resources department,

assign the task of developing the recognition menu to the safety committee. However, if the safety committee is not broadly representative, form an ad hoc committee of the safety committee and add as many members as necessary to ensure that all major functional areas in your organization are represented. Make sure that a representative of the human resources department is on the ad hoc committee.

Step 2: *Create a Master List of Potential Recognition Rewards*

I always recommend that organizations prepare for this step by purchasing copies of the book *1001 Ways to Reward Employees*, by Bob Nelson. This book contains, as the title implies, more than 1,000 ideas for rewards that can be given to employees who have been selected for recognition. Approximately half of the rewards in the book are nonmonetary in nature. Remember that the recognition menu should contain only nonmonetary rewards—no raises, bonuses, or incentive dollars. Of course, the term nonmonetary does not mean that the rewards in question will not cost the organization money. For example, a reward of a gift certificate to a local restaurant has an associated cost. *Nonmonetary* in the current context means that the employees in question receive no money. Each member of the committee is asked to review Bob Nelson's book and pull from it as many potential recognition rewards as possible. Members should use their intuition and experience to decide if the chosen items are appropriate for your organization. A draft master list of potential rewards is created from the list compiled by each committee member. A good number for the draft master list is 50 to 60 potential rewards. The draft list is submitted to higher management for approval.

Step 3: *Finalize the Master List of Recognition Rewards*

It is not uncommon for higher management to strike several potential recognition rewards from the draft master list. This is one of the reasons I recommend that the draft list submitted to higher management contain between 50 and 60 items. With this many, the approved list will typically contain between 40 and 50 items. This list is going to be narrowed down to the 25 most attractive items by your organization's employees in this step. The process works like this. The approved master list is provided to all employees. This can be accomplished electronically using an Internet survey engine such as "Survey Monkey" or manually by using supervisors as the key points of contact or a combination of these approaches depending on the exigencies of your organization. All personnel are asked to choose the top 25 items on the master list on the basis of how appealing they find each item as a reinforcer and motivator. The top 25 items, as selected by your organization's personnel, make up the final list.

Using the Recognition Menu

Once the recognition menu has been finalized, it is ready for use as soon as your organization selects its next cycle of individuals or teams for recognition (e.g., Employee of the Month, Quarter, or Year or any other performance-oriented recognition). Assume that an employee has been selected as "Outstanding Performer of the Month" (an award that should include safety-related performance); rather than receive a plaque, framed certificate, parking space, or whatever recipients of this recognition have been given in the past, the individual in question is allowed to select an item from the recognition menu. The recipient might actually select a parking space or a plaque (provided those items are part of the menu) or he or she might select something else. The point is that whatever item is selected, it represents the employee's choice rather than a one-size-fits-all item that might not appeal to the recipient.

Updating the Recognition Menu Periodically

The recognition menu must be updated periodically or it will lose its effectiveness. I recommend updating the menu no less than every two years. This involves repeating the process used for developing the menu in the first place with one addition. When updating the menu, all items on the draft master list that are included in the current recognition menu should be highlighted in some way and notated to show how often they have been selected by employees as part of the recognition process. Popular items should stay on the menu. Others may be changed, on the basis of the level of interest expressed when the draft list is circulated among employees, every time the menu is updated.

Categorizing the Items in the Recognition Menu

An employee who is recognized as employee of the year should receive a higher-value reward than one who is recognized as employee of the quarter. An employee who is recognized as employee of the quarter should receive a higher-value reward than one who is recognized as employee of the month. An effective way to handle this situation without developing more than one recognition menu is to stratify the potential rewards by level of monetary value (e.g., a gift certificate for $100 would be categorized at a higher level than two $7 movie tickets). I once helped an organization develop a recognition menu that contained 30 items arranged in three different categories: Levels 1, 2, and 3. Level 1 rewards had the highest monetary value and were for employees being recognized as "Outstanding Performer of the Year." Level 2 rewards were for employees being recognized as "Performer of the Quarter." Level 3 rewards went to employees being recognized as "Performer of the Month."

Our intent in developing a stratified menu was to avoid the confusion we thought would result from circulating several different draft lists among employees each year, one draft for each level of reward. Our stratified menu served this purpose well, but it also had a salutary effect we had not anticipated. When employees saw the types of rewards that accompanied Levels 2 and 3, they became even more motivated to perform well in all categories measured, including safety. Employees began to put forth extra effort to get to Level 1. Level 1 recipients immediately set a goal of moving up to Level 2, and Level 2 recipients set a goal of moving up to Level 3.

The menu approach for recognizing high performers can be an effective way to reinforce the types of attitudes and practices your organization expects from its personnel. The key to success from the perspective of safety is to ensure that safety-related criteria have a high priority when selecting employees for special recognition. The safety committee is responsible for making sure that safety-related criteria are part of the selection process and that they are given an appropriate priority among the other criteria.

SAFETY-FIRST CORPORATE PROFILE
Gulf Power Company

Gulf Power Company is an investor-owned electric utility company that is part of the Southern Company. The Southern Company consists of, among other holdings, several electric utility companies, including Mississippi Power, Alabama Power, Georgia Power, and Gulf Power. Gulf Power Company serves several western counties along the Gulf of Mexico in Florida's Panhandle. The company's main office is in Pensacola, Florida. Gulf Power Company is in the business of generating and distributing electric power, both potentially dangerous undertakings. In fact, according to the NSC, there are approximately 12,000 electricity-related accidents in the United States every year. Of these, more than 500 result in deaths. Consequently, safety must be a high priority at Gulf Power Company—and it is. Gulf Power Company's overall goal for safety is "target zero" or a goal of zero accidents annually.

Gulf Power Company has a comprehensive safety program designed to achieve "target zero" that includes orienting new personnel, training, and mentoring. But the company concerns itself with more than just the safety of its personnel. It is also reaches out to its customers and provides them with assistance to protect themselves from electricity-related accidents and incidents. Gulf Power provides electric safety information, assistance, and education to its customers in the following areas:

- Power lines
- Around the house
- Power tools and cords
- Machinery and power lines
- Circuits, wiring, and ground fault circuit interrupters (GFCIs)
- Electrical Fires
- Call before you dig
- Learning Power website
- "Safety City"

Gulf Power Company's various programs for informing and educating its customers about how to use electricity safely are—when taken together—an excellent example for other organizations that provide products that are potentially dangerous. Protecting your personnel is important, but don't forget about your customers.

Source: http://www.gulfpower.com

COLLEAGUE-TO-COLLEAGUE DISCUSSION CASES

CASE 1: Let's Establish an Awards Program for Safety Performance

The safety committee at ASD, Inc., has made excellent progress toward establishing a safety-first corporate culture. Today's meeting concerns revising the company's approach to selecting employees for recognition awards so that safety-related criteria become part of the mix. One member of the team has suggested establishing a separate program for making recognition awards based solely on safety-related criteria. This program would be separate and apart from the company's established recognition program. Another member of the safety committee is not sure that a separate program for safety awards is a good idea. She wants to see ASD, Inc., revise its established selection methods to include safety-related criteria.

Frank Asherton—ASD, Inc.'s, safety director—is not sure what to do. He can see some merit in both proposals. On the one hand, establishing a separate recognition program strictly for safety-related awards would be the easiest thing to do. The safety committee could establish such a program itself with nothing more than the approval of higher management. There would be no need to involve the human resources department or to revise an existing set of criteria or to persuade supervisors and other management personnel to buy into safety-related criteria. On the other hand, Asherton knows that safety will never be a high priority in the company as long as it is treated as a stand-alone entity that is just "tacked on" for good measure.

Asherton can tell there is no clear consensus among the safety committee's members on this issue. Most of them are like him in that they can see merit in both proposals. After listening to the debate in the safety committee for more than 30 minutes with no consensus in sight, Asherton decides to adjourn the meeting so he can research the issue further and be better prepared to discuss it at a future date.

Discussion Questions

1. Have you or your colleagues ever worked in an organization that included safety-related criteria when selecting personnel for recognition awards? Have you or your colleagues ever worked in an organization that had a separate recognition program solely for safety awards? Discuss the pros and cons of each situation.

2. If Frank Asherton asked for your advice concerning how to resolve his current dilemma, what would you tell him?

CASE 2: What Is This Menu Approach I Keep Hearing About?

"I read an article last night in a professional journal about something called the *menu approach* for recognizing employees for outstanding performance." This statement was made by Charla Katle, a member of ERT, Inc.'s, safety committee. Katle explained that, according to the article, today's employees are so accustomed to having options that the traditional one-size-fits-all approach to giving recognition awards does not work well any more. "According to this article I read, the current generation of employees grew up with 200 television channels; cell phones with hundreds of service options; the ability to shop online, by telephone, or in person; and many other kinds of options in their lives. Then they come to work for us, and our recognition system has no options." "That's not good news for us," said John Bellon. Bellon is ERT, Inc.'s, safety director. "Ours is definitely a one-size-fits-all recognition program."

Katle recommended that the safety committee take responsibility for approaching the human resources department and higher management with the idea of revising ERT, Inc.'s, recognition system. She thought the menu approach would be better for the company. Bellon agreed with her, but was not sure how to proceed. The safety committee discussed next steps, but had arrived at no definitive plan when Bellon decided to adjourn the meeting to give members more time to think about how to proceed.

Discussion Questions

1. Have you or your colleagues ever worked in a situation in which you needed to win the support of another department to implement an initiative that would be good for safety? If so, discuss the situation and how you handled it.

2. If John Bellon called you and asked for advice on how to proceed, what would you tell him?

Key Terms and Concepts

Before leaving this chapter, make sure you understand the following key terms and concepts and can accurately explain them to people who are not safety professionals:

Rewards

Recognition

Nonmonetary reinforcement

Reward-and-recognition system

Integrated reinforcement system

Menu approach for recognizing personnel

Master list of potential recognition

Review Questions

Before leaving this chapter, make sure you can accurately and comprehensively, but succinctly, answer the following review questions:

1. Explain the rationale for reinforcing safety-first attitudes and practices.
2. Explain how rewards and recognition can be used to reinforce the expected safety-first attitudes and practices.
3. Which is more likely to work best as a reinforcer of safety-first attitudes and practices: monetary rewards or recognition? Why?
4. Explain the rationale for adopting an integrated reinforcement program.
5. Explain the menu approach for recognizing high-performing employees.

Application Project

Establishing an integrated recognition program is another of those challenges that requires the safety committee to work with human resource personnel and to seek approval from higher management. However, ensuring that all aspects of safety and health are fully integrated into your organization's corporate culture will make the effort worthwhile. The safe and healthy way will become the "normal and expected way" only when safety is a fully integrated component of your organization's operations.

Consequently, the project for this chapter is to develop a comprehensive plan for (1) establishing a fully integrated recognition system that includes safety- and health-related selection criteria and gives them an appropriately high priority and (2) adopting the menu approach for recognizing high-performing personnel.

Endnotes

1. Anheuser-Busch Companies. Retrieved May 2008 from http://www.abenvironment.com

2. As quoted in *1001 Ways to Reward Employees*, by Bob Nelson (New York: Workman Publishing Company, 1994), 19.

Chapter 10

Periodically Assess the Safety Aspects of Your Organization's Corporate Culture

Major Topics

- Review of assessment basics
- Sample assessment instrument
- Conducting the assessment using a survey instrument
- Plan for improvement
- Implement the improvement plan
- Monitor and adjust

At this point in the establishment of a safety-first corporate culture, the standard management model of *assess-plan-implement-monitor-adjust* comes into play. The first nine steps (Chapters 1 to 9) are designed to establish a safety-first corporate culture in your organization. This step asks the question, how are we doing? Then, on the basis of the answer to this question, a plan for improvement is developed and implemented. The effectiveness of the implementation is monitored, and adjustments are made as appropriate. This application of the standard management model continues forever, which, in turn, ensures continual improvement—an important aspect of a safety-first corporate culture.

Do we have a safety-first corporate culture in our organization? What are the attitudes of our personnel toward safety? Are our personnel committed to safe work practices? No matter how hard you have worked to establish a safety-first corporate culture in your organization and no matter how effective your efforts appear to have been, don't make the mistake of thinking, "We are finally there—now we can just sit back and enjoy the results." Once established, a safety-first corporate culture must be maintained. *Establishing* a safety-first corporate culture can be approached as a project—a major organization-wide effort that has a beginning and an end. However, *maintaining* a safety-first corporate culture is a process—not a project—and it continues forever.

An important step in maintaining a safety-first corporate culture is to periodically assess the safety aspects of your organization's corporate culture. Then, on the basis of the results of the assessment, develop a plan for making any needed improvements, implement the plan, monitor results, and adjust as necessary. This is a process that goes on forever, which is why maintaining a safety-first corporate culture must be viewed as a process rather than a project. As with all of the steps explained in this book, your organization's safety professionals and the safety committee must play the key facilitating role in this step. However, as with the other steps in the model, it will be necessary to work closely with a broad cross-section of executives, managers, and supervisors throughout your organization in implementing this step.

Review of Assessment Basics

An organizational assessment is a measure of where your organization is at a specific point in time compared with where it wants to be; in this case, as the question relates to safety. The assessment process recommended in this chapter is best carried out by your organization's safety committee. When preparing to conduct an assessment, the safety committee should consider the following foundational factors: why, what, whom, when, and how.[1] A good assessment instrument (next section) will help your organization answer these types of questions as they relate to a safety-first corporate culture.

Why Factor

The safety committee should begin its discussion of the assessment by asking the following question: Why are we doing this? The answer to this question may be obvious to you, but do not make the mistake of assuming that it is obvious to everyone. First, it is important to ensure that all members of the safety committee understand the reason for conducting the assessment. Second, it is important that all members of the safety committee be able to articulate to executives, managers, supervisors, and employees why the assessment is being conducted. Remember, as with every step in the model for establishing a safety-first corporate culture, buy-in will be needed from key personnel throughout the organization. Make sure that everyone on the safety committee understands why the assessment is being undertaken and can articulate its purpose in terms that are both succinct and persuasive.

The message that safety committee members must be able to help others understand is this: *The assessment is being conducted to determine the extent to which the organization has a safety-first corporate culture.* The assessment will give decision makers valuable information about organizational strengths and weaknesses related to the safety aspects of the corporate culture. On the basis of the results of the assessment, decision makers will be better able to develop an informed plan for closing any gaps between the actual and the desired state of the corporate culture relating to safety.

What Factor

What is the assessment supposed to determine? The big picture answer to this question is that the assessment should determine the extent to which the tacit assumptions, beliefs, values, attitudes, expectations, and behaviors that are widely shared and accepted in the organization support the maintenance of a safe and healthy workplace. More specifically, the assessment should determine how committed managers, supervisors, and employees are to workplace safety. It should answer the question, what is the level of our organization's commitment to a safe and healthy work environment?

Whom Factor

This question has at least three components: (1) For whom is the assessment being conducted? (2) Who will participate in the assessment? and (3) Who will conduct the assessment? When assessing the safety aspects of your organization's corporate culture, the assessment is being conducted on behalf of senior management. This is why gaining the commitment of your organization's CEO and executive management team to a safety-first corporate culture is so important. Gaining executive-level commitment was covered in Chapter 2.

When assessing the safety-related aspects of the corporate culture, it is best to involve everyone. Unless your organization is so large as to make this impossible, all personnel should be encouraged to participate. Larger organizations might have to use a representative sample of all personnel, but having everyone participate is the preferred approach. The broader the participation, the more valid the data collected are likely to be.

When Factor

The natural inclination is to want to conduct the assessment right away, and *the sooner the better* is generally a good rule of thumb with assessments. However, timing is an important factor in determining the effectiveness of an assessment. Consequently, the following guidelines should be observed when establishing the time frame for conducting an organization-wide assessment:

- Avoid those times of year when employees typically take their vacations.
- Do not let circumstances rush you into doing a less than credible assessment.
- Make the period of time in which to respond to the assessment instrument long enough so that personnel can work on it off and on as their workload allows.
- If the assessment methodology involves interviews and/or focus groups, conduct the sessions before lunch when personnel are more likely to be fresh rather than at the end of the day when they are fatigued.
- If the assessment methodology involves completing a survey instrument, make sure personnel have plenty of time to fill it out and that they can put the instrument aside or save it electronically and come back to it.
- Remember that everything will probably take longer than you think it will, and develop your schedule for conducting the assessment accordingly. Do the same for compiling and analyzing the results.

How Factor

The three most common assessment methods are interviews, focus groups, and surveys. Most safety professionals are experienced in using interviews to solicit the facts following workplace accidents and incidents. The same techniques can be used to identify the perceptions of people concerning your organization's corporate culture as it relates to safety. However, interviews are time consuming and lack the benefit of allowing participants to respond without attribution. This is a serious flaw because anonymity typically promotes more honest responses concerning culture-related perceptions.

Focus groups are an excellent way to solicit input from representative groups of personnel in an organization. Think of focus groups as interviews with groups rather than individuals. Focus groups allow for a broader base of participation than interviews, but they require special skills on the part of the facilitator. A facilitator must be well versed in such group-dynamics techniques as drawing out reluctant participants and reining in those who would dominate rather than participate.

In addition, to keep the discussion of focus groups on track and moving in the right direction, as opposed to wandering off down rabbit trails, it is necessary to develop what is sometimes called a "strawman." Focus groups that begin with a clean sheet of paper often overlook critical issues that need to be part of the discussion. A strawman is a draft list containing suggested issues for discussion. The strawman is used to plant ideas and to get the ball rolling, not as a final document to simply be accepted by the group.

Surveys require the most preparation, but they have the advantages of (1) allowing the broadest possible participation and (2) permitting participants to provide their input anonymously. Further, with the aid of computer technology, surveys can be distributed, completed, and compiled electronically. These benefits are why I recommend using the survey as the main tool for conducting the assessment in this step, but with a caveat. The focus group approach should be used by the safety committee for developing a draft of the survey instrument, and the interview method should be used for soliciting input from executive-level managers concerning the survey instrument and for gaining their approval of the instrument.

Sample Assessment Instrument

The assessment instrument in Figure 10.1 is provided as an example the safety committee can use for developing an instrument tailored specifically for your organization. It contains fewer than 20 criteria—an important characteristic for this kind of survey. The instrument used in an organization-wide survey should meet a number of criteria including the following:

- Comprehensive enough to adequately serve its purpose
- Brief enough so that it can be completed in approximately 10 minutes or less
- Simple enough so that it can be understood by personnel at all levels
- Able to be administered electronically or manually
- Quantifiable electronically or manually

The example in Figure 10.1 meets all of these criteria. However, rather than just adopt Figure 10.1 verbatim, the safety committee should use it as a strawman for conducting a focus group. All members of the safety committee should participate in the focus group as well as additional personnel to ensure that all departments in the organization are represented.

Assessment Instrument

Safety-Related Aspects of Our Corporate Culture

The purpose of this instrument is to assess the extent to which our organization has a *safety-first corporate culture*. In organizations with a safety-first corporate culture, the tacit assumptions, beliefs, values, attitudes, expectations, and behaviors that are widely shared and accepted by personnel at all levels support the maintenance of a safe and healthy work environment. Please assist the safety committee in assessing the safety-related aspects of our corporate culture by indicating your responses to the following statements using the code provided below:

4 = Completely true
3 = Somewhat true
2 = Somewhat false
1 = Completely false
X = Do not know or not applicable

Your classification: _____Management _____ Supervisor _____Employee

_____ Top management is committed to a safe and healthy workplace.
_____ Providing a safe and healthy work environment is viewed by top management as part of our organization's corporate social responsibility.
_____ Safety and health are high priorities in our organization.
_____ Key decision makers in our organization view a safe and healthy work environment as a competitive advantage.
_____ Our organization views employees as valuable assets to be protected from workplace hazards.
_____ Our orientation for new personnel stresses the importance of workplace safety and health.
_____ Our mentoring program helps new personnel develop safe and healthy attitudes and work practices.
_____ Our team-building efforts include the development of safety-first attitudes and work practices.
_____ When employees are recognized for outstanding performance, their attitudes and practices relating to safety are high-priority considerations.
_____ Safety and health are important considerations when decisions are made in our organization.
_____ Managers, supervisors, and employees view a safe and healthy work environment as the most conducive environment for peak performance and continual improvement.
_____ Our organization makes it clear that the safe way is the right way and the expected way in all instances.
_____ Employees are encouraged to speak up when they have concerns about safety or health issues.
_____ Daily monitoring by supervisors includes safety–related attitudes and practices.
_____ Internal peer pressure supports doing things the safe way.
_____ Our organization's unwritten rules support doing things the safe way.
_____ All personnel in our organization receive the training and retraining they need to do their part in maintaining a safe and healthy work environment.

Comments:

FIGURE 10.1 The safety-related aspects of your organization's corporate culture should be assessed periodically.

Criteria relating specifically to your organization may be added to the instrument, but be cautious of eliminating any criteria unless they simply do not pertain. The criteria contained in Figure 10.1 will apply in most cases. Each criterion is explained in the remainder of this section.

Top Management Is Committed to a Safe and Healthy Workplace

In safety-first organizations, top management is committed to a safe and healthy workplace. The importance of top management's commitment to safety has been stressed throughout this book. In addition, various ways that this commitment can be translated into action have been summarized. But what are the perceptions of your organization's personnel concerning top management's commitment? Do the personnel in your organization believe that top management is committed to a safe and healthy workplace? The perception question is important. No matter what top management does to ensure a safe work environment, there is still a problem if a perception of commitment does not exist throughout the organization.

The attitudes of personnel at all levels in your organization toward safety will tend to reflect their perception of top management's commitment to safety. If the survey reveals that, even in spite of excellent efforts on the part of top management, a perception of commitment does not yet exist, there is still work to do. It is not enough for executives to be committed to a safety-first corporate culture. They must communicate that commitment to personnel in ways that create a positive perception that is broadly shared.

Providing a Safe and Healthy Work Environment Is Viewed by Top Management as Part of Our Organization's Corporate Social Responsibility

In safety-first organizations, providing a safe and healthy work environment is viewed by top management as part of the organization's corporate social responsibility. If your organization's top management team adopts a statement of corporate social responsibility—which it should—workplace safety and health should be included in it. This is an important step in making sure that safety and health become fully integrated into your organization's policies, processes, and procedures. This criterion is to remind all personnel that (1) top management needs to adopt a statement of corporate social responsibility, (2) safety and health need to be part of this statement, and (3) this statement needs to be effectively communicated to all personnel in your organization.

Safety and Health Are High Priorities in Our Organization

In safety-first organizations, safety and health are high priorities. This criterion might seem redundant since the first two criteria concern top management's commitment to safety and health (i.e., how they prioritize it). The redundancy is intentional and has a purpose. This criterion is included to gauge the perceptions of personnel as to the reality of the priority your organization gives to safety and health. Corporate social responsibility statements that include safety are important. However, do managers, supervisors, and employees just talk about safety, or do they actually reinforce their words with action? This question gets at the *talk versus action* concern.

Key Decision Makers in Our Organization View a Safe and Healthy Work Environment as a Competitive Advantage

In safety-first organizations, key decision makers view a safe and healthy work environment as a competitive advantage. Organizations succeed by establishing and exploiting characteristics and capabilities that allow them to outperform the competition. These competitive advantages are included in the organization's strategic plan. It was established early in this book that a safe and healthy workplace can be a competitive advantage because it provides an environment that promotes peak performance and continual improvement. Do key decision makers in your organization view safety and health as a competitive advantage? If they do, the organization's strategic plan will say so. Further, personnel at all levels should know that safety and health are viewed in this way. Do personnel in your organization believe that key decision makers view a safe and healthy work environment as a competitive advantage, or do they think that decision makers just give lip service to safety and ignore it when push comes to shove? This criterion or one similar to it will answer these types of questions.

Our Organization Views Employees as Valuable Assets to be Protected from Workplace Hazards

In safety-first organizations, employees are viewed by management not just as labor, but as valuable assets that are critical to the performance, competitiveness, and success of the organization, assets that, because of their value to the organization, must be protected from workplace hazards. Organizations that view their personnel as assets do what is necessary to keep them safe, healthy, and performing at peak levels. To do otherwise is to undermine the organization's competitiveness. Does your organization view its personnel as assets? More to the point, do your personnel feel that their organization views them as assets? If so, this perspective will be apparent in daily operations, interactions, and decisions. If not, the opposite perspective will be apparent. This criterion or one similar to it will answer these types of questions.

SAFETY-FIRST FACT

Upper Management Must Understand the Importance of Safety

DynMcDermott Petroleum Operations Company manages the strategic petroleum reserve for the U.S. Department of Energy and has an award-winning safety program. According to the company's safety and health manager, Suzanne Broussard, DynMcDermott's success in the area of safety can be attributed to the fact that upper management understands the importance of workplace safety.

Source: Retrieved from http://www.nsc.org/news/nr110606.htm on March 11, 2008.

Our Orientation for New Personnel Stresses the Importance of Workplace Safety and Health

In safety-first organizations, the orientation is viewed as an opportunity to get new employees started on the right foot in several critical areas, including, of course, safety. Since it is important to ensure from the outset that new employees adopt safety-first attitudes and practices, safety must be integrated as appropriate into all aspects of the orientation. Do employees feel that safety and health were stressed appropriately during their orientation? Could the safety aspects of the orientation be strengthened? This criterion or one similar to it will answer these types of questions.

Our Mentoring Program Helps New Personnel Develop Safe and Healthy Attitudes and Practices

In safety-first organizations, new employees are provided mentors to help them develop the attitudes and practices that will lead to consistent peak performance and continual improvement. Mentors help guide new employees through the maze of uncertainty that goes with being a "rookie" in an organization. A good mentoring program includes the development of safety-first attitudes and work practices. Do employees feel that their mentors helped them develop a safety-first attitude and work practices? This criterion or one similar to it will answer this type of question.

Our Team-Building Efforts Include the Development of Safety-First Attitudes and Practices

In safety-first organizations, safety and health are fully integrated aspects of team-building efforts. Chapter 7 was devoted to the various ways safety and health can be integrated into team building. How effectively does your organization accomplish this goal? Do employees believe that safety-first attitudes and work practices are developed as part of your organization's team-building efforts? Do your organization's team-building efforts need to be revised and improved? This criterion or one similar to it will answer these types of questions.

When Employees Are Recognized for Outstanding Performance, Their Attitudes and Practices Relating to Safety Are High-Priority Considerations

In safety-first organizations, awards that recognize outstanding performance cannot be earned without outstanding performance in the area of safety. Chapter 9 was devoted to using recognition and other methods to reinforce safety-first attitudes and practices. Do the criteria used for singling out personnel for performance awards include safety-related criteria? Do personnel in your organization believe they must work safely and maintain positive attitudes toward safety in order to be recognized for outstanding performance? Do they believe that safety is a high-priority consideration when selecting employees for recognition awards? This criterion or one similar to it will answer these types of questions.

Safety and Health Are Important Considerations When Decisions Are Made in Our Organization

In safety-first organizations, decision makers at all levels consider the safety-related ramifications of the decisions they make. Some decisions have no safety-related ramifications, but many do. A decision that appears to solve a problem, gain an advantage, or create a beneficial result will sometimes look different when viewed from the perspective of workplace safety and health. For example, cutting the training budget during difficult financial times might appear on the surface to save badly needed dollars. However, when basic safety training is cut from the budget, the long-term costs are likely to outweigh the short-term savings. Decision makers who are committed to safety understand this and act accordingly. Do decision makers in your organization consider the safety-related ramifications when making decisions? This criterion or one similar to it will answer this type question.

Managers, Supervisors, and Employees View a Safe and Healthy Work Environment as the Most Conducive Environment for Peak Performance and Continual Improvement

In safety-first organizations, a safe and healthy work environment is viewed by decision makers as a competitive advantage because it encourages and enables peak performance and continual improvement. However, it is one thing to make such a statement in your organization's strategic plan, and it is quite another to transform this philosophical ideal into everyday practice. Do managers, supervisors, and employees view a safe and healthy work environment as the most conducive environment for peak performance and continual improvement? This criterion or one similar to it will answer this type of question.

Our Organization Makes It Clear That the Safe Way Is the Right and Expected Way in All Instances

While in Marine Corps boot camp at Parris Island many years ago, my fellow recruits and I learned from the outset that there were only two ways of doing anything: the Marine Corps way and the wrong way. We also soon learned that the wrong way was not an option. To even attempt to do something "our way" or any way not prescribed in the Marine Corps' Manual for Recruits was unthinkable. This is how it is in safety-first organizations. The only right way is the safe way, and the safe way is "our way." This philosophy must be reinforced by managers, supervisors, and employees every day and in every instance. The philosophy that *anything is good as long as it works* is not acceptable in a safety-first organization. Do personnel at all levels in your organization make it clear that the safe way is the right way, or do they overlook safety violations as long as no one gets hurt? This criterion or one similar to it will provide the answer to this type of question.

Employees Are Encouraged to Speak Up When They Have Concerns about Safety or Health Issues

In safety-first organizations, employees are encouraged to speak up when they have concerns about safety and health issues. But they are not just encouraged, they are provided opportunities for doing so. Further, they are protected from "kickback" in any form when they speak up about safety concerns. Employees are closer to the everyday work of your organization than supervisors and managers. Consequently, they will sometimes see things that others more removed from the work might overlook. When this happens, employees must have avenues for making their concerns known, and they must be encouraged to use them. Further, they must know that their concerns will be heard, given careful consideration, and acted on in an appropriate and timely manner without even a hint of the *shoot the messenger syndrome*. In fact, they must feel as if they will be thanked for raising the issue of potentially hazardous conditions. Do employees in your organization believe they are encouraged to speak up when they have safety-related concerns? This criterion or one similar to it will provide an answer to this type of question.

Daily Monitoring by Supervisors Includes Safety-Related Attitudes and Practices

In high-performance organizations, supervisors monitor the work attitudes and practices of their direct reports continually. In high-performance organizations that have a safety-first corporate culture, this monitoring includes safety-related attitudes and practices. It is important to ensure that unsatisfactory attitudes and practices are not allowed to persist, lest they become habitual and spread like a cancer in your organization. Further, it is equally important to reinforce positive work attitudes and practices in real time. Do employees believe that their supervisors include safety-related attitudes and practices in their daily monitoring? This criterion or one similar to it will answer this type of question.

Internal Peer Pressure Supports Doing Things the Safe Way

In safety-first organizations, peer pressure among employees supports doing things the safe way. A common problem in organizations that have not established a safety-first corporate culture is that employees often work one way when the "boss" is looking and another way when he or she is not. In such organizations, employees will look the other way when their peers cut corners and ignore safety precautions. In fact, it is not uncommon for employees in these organizations to actually encourage their peers to ignore safety rules and regulations. How do your organization's personnel work when not being observed by someone in a position of authority? Do your organization's employees hold each other accountable for safe work practices, or do they look the other way when violations occur? This criterion or one similar to it will answer these kinds of questions.

Our Organization's Unwritten Rules Support Doing Things the Safe Way

In safety-first organizations, personnel can assume that the safe way is not just the right way, but the expected way. There is no question. When production falls behind schedule, employees know that cutting corners on safety is not how they are to catch up. Safety-first companies are typically high-performance organizations, and high-performance organizations meet their deadlines. However, ignoring safety rules and regulations is not one of their catch-up strategies. There are always other options available for getting production schedules back on track. Do your organization's personnel understand that the safe way is the expected way even when the situation in question is not covered by specific written rules and regulations? This criterion or one similar to it will provide the answer to this question.

All Personnel in Our Organization Receive the Training or Retraining They Need to Do Their Part in Maintaining a Safe and Healthy Work Environment

In safety-first organizations, training is at high priority. If personnel are expected to work safely, they must be taught how. High-performance organizations never assume that their personnel know how to do what is expected of them. Rather, they provide training to make sure. Further, safety-related training is not just a one-time undertaking. It is provided on a regular basis to ensure that all personnel are kept up to date with new methods, hazards, regulations, and standards. The critical aspect of this criterion is found in the words "They need to do their part."

Often decision makers in organizations think they know what kinds of training are needed by their personnel, and, in most cases, the needs are obvious, but not always. Sometimes certain personnel will need a different kind of training than is obvious to supervisors and managers. This criterion or one similar to it will let decision makers know if there are safety training needs that are not being met by your organization's safety training program.

Conducting the Assessment Using a Survey Instrument

Once the survey instrument has been finalized, it can be used as the principal tool in conducting an organization-wide assessment. This section explains a number of strategies that can improve the quality of the survey process while also keeping it simple and manageable.

- Before distributing the survey instrument, communicate with all personnel organization-wide about it. Let them know the purpose of the survey (as stated on the survey instrument). In addition, explain that their frank and objective participation is not just desired, but will be the key to the effectiveness of the process. This message is best communicated by your organization's CEO or another senior manager on the CEO's behalf. The important point here is to let all personnel know that the survey has the approval and support of the CEO and top management. This message of support at the highest level should then be reinforced by managers and supervisors throughout the organization.

• Explain that the purpose in asking personnel to indicate their status in the organization (i.e., manager, supervisor, or employee) is to allow the safety committee to stratify the results of the survey and, in turn, analyze them more effectively. For example, if managers rate a certain criterion as being "completely true," but employees rate it as "somewhat false," the safety committee will know there is a perception gap related to this criterion.

• Make sure that all personnel understand that their responses are confidential, that they cannot be traced back to them, and that there is no desire to do so. Ratings entered by personnel on the survey instrument are submitted without attribution. Explain that asking personnel to classify themselves is simply a mechanism for determining if personnel at different levels in the organization have different perceptions (a common problem in organizations).

• Encourage participants to make comments at the end of the survey instrument, especially comments that might help improve the safety and health of the work environment.

• Unless your organization is too small to have one, the safety committee should work with the information technology department to load the survey instrument on the Web so it can be completed online and the results compiled electronically. If your organization does not have an information technology department, I recommend using a Web-based tool such as Survey Monkey (http://www.surveymonkey.com). In setting up the survey online, make sure that the design of the compilation component rules out including criteria marked "X" when computing the mean score for each criterion. Any criterion marked with an "X" should not be counted when determining the organization-wide average or mean score for that criterion.

• Analyze the compiled results—the mean score for each criterion—to determine which criteria have unacceptably low mean scores as well as where there are differences between the scores of management personnel and employees. Recommending a cutoff score for determining what is "unacceptable" is the responsibility of the safety committee, with final approval coming from higher management. However, any mean score below three should certainly be considered unacceptable. The safety committee may wish to recommend an even higher cutoff score than three.

• Extract from the organization-wide results all survey criteria that have a mean score below the cutoff point approved by higher management. These criteria are then prioritized and used in the next step in the process: developing a plan for improvement. Before developing the improvement plan, a decision must be made. Does your organization want to prioritize the improvement criteria according to mean score (e.g., lowest score = highest priority), or would it rather prioritize by relative importance to the organization? For example, assume that the criteria "Our mentoring program helps new personnel . . ." and "Top management is committed to a safe . . ." both receive an unacceptably low mean score. To complicate things even further, employee mean scores concerning top management's commitment are lower than those of executives. The overall mean score of the "mentoring" criterion is actually slightly lower than that of the "commitment" criterion. However, because of the fundamental problems associated with a lack of commitment from top management or, at the very least, different perceptions between management and employees about this commitment, the safety committee might want to recommend giving the "commitment" criterion a higher priority for improvement. All criteria that achieve a mean score lower than the cutoff point must be targets for improvement. However, limited resources and the nature of the shortcoming can sometimes force an organization to prioritize according to factors other than just mean score. In fact, I recommend prioritizing according to perceived relative importance rather mean scores, but with a caveat. If there are improvements that can be made quickly and easily, go ahead and make them. This approach is sometimes referred to as "picking the low-hanging fruit." The rationale for this approach is that success breeds success and builds positive momentum. When managers and employees can see evidence that progress is being made toward achieving improvement goals, they are more likely to jump on the safety-first bandwagon and give their wholehearted support to the effort.

Once the assessment of the safety aspects of your organization's corporate culture has been completed and the criteria in need of improvement have been prioritized, the next step is to develop a plan for improvement. The improvements in question are those that need to be made if your organization is to have a safety-first corporate culture.

Plan for Improvement

The first step in developing a plan for improvement is for the safety committee to compare the areas that need improvement with the various steps presented in this book. To review, these steps are as follows:

1. Gain commitment to a safety-first corporate culture.
2. Expect appropriate safety-first attitudes and practices.
3. Role model the expected safety-first attitudes and practices.
4. Orient personnel to the expected safety-first attitudes and practices.
5. Mentor personnel on the expected safety-first attitudes and practices.
6. Train personnel on the expected safety-first attitudes and practices.
7. Make safety part of team building.
8. Monitor and evaluate safety-first attitudes and practices.
9. Reinforce safety-first attitudes and practices.
10. Periodically assess the safety aspects of the corporate culture.

Planned improvements are grouped according to the specific strategy on this list that they relate most closely to. For example, assume the "mentoring" criterion in the survey received an unacceptable mean score. After discussing the matter and looking into it further, the safety committee decides that (1) the process for selecting mentors must be strengthened, and (2) the personnel selected to be mentors must receive more and better training. These planned improvements would be listed under the mentoring strategy, which would be a major heading in the improvement plan.

As an example of how the improvement plan is developed by the safety committee, consider the following scenario. Assume that the mean score for the "recognition" criterion in the survey is unacceptably low. This criterion reads, *"When employees are recognized for outstanding performance, their attitudes and practices relating to safety are high-priority considerations."* Since recognition is an important form of positive reinforcement, this criterion falls under Step 9: *Reinforce safety-first attitudes and practices.* The safety committee, after looking into the matter, determines that (1) a high percentage of supervisors are not considering safety attitudes and practices when recommending employees for recognition, and (2) there have been instances in which personnel who have been reprimanded for safety violations have actually received recognition awards on the basis of other aspects of their performance.

After discussing this situation, the safety committee decides to (1) strengthen the safety aspects of the organization's training program for supervisors—especially the part where supervisors learn how to choose personnel for recognition awards and (2) have one-on-one conferences with supervisors who have recommended poor safety performers for performance awards. In the plan for improvement, this entry would appear as follows:

REINFORCE SAFETY-FIRST ATTITUDES AND PRACTICES

1. Strengthen the safety aspects of our organization's training program for supervisors—especially the part where supervisors learn how to choose personnel for recognition awards.

2. Conduct one-on-one conferences with supervisors who have recommended poor safety performers for performance awards.

Implement the Improvement Plan

Because so much of what must be done to continually improve the safety aspects of an organization's corporate culture falls outside of their direct authority, implementing the improvement plan will require safety professionals and the safety committee to work closely with a broad cross-section of other management personnel. The first step is to seek approval from higher management for the improvement plan. It cannot be implemented without executive-level approval. Of course, this has been the case with all of the steps presented in this book and is probably the most difficult ongoing challenge when attempting to establish and maintain a safety-first corporate culture. But in this step, the challenge can become even more pronounced.

For example, executives, managers, and supervisors will not always welcome the news that some performance criterion for which they are responsible needs improvement or that people who report to them need to improve their performance. This is an understandable reaction. However, until executives, managers, and supervisors accept the results of the organization-wide survey in a positive manner and are willing to do their respective parts to ensure successful implementation of the improvement plan, a safety-first corporate culture does not yet exist. Consequently, it is important that safety professionals and safety committees persevere in this step and continue to diplomatically apply an appropriately balanced "carrot and stick" program.

Monitor and Adjust

Once the improvement plan is approved by higher management, the various strategies in it can be assigned to specific individuals in your organization and time frames for their completion worked out. This is an important step because it builds the concept of accountability into the implementation process. Monitoring the implementation process is the responsibility of the safety committee, but carrying out the various improvement strategies in the plan will require the involvement of a broad cross-section of personnel. This is why it is so important that every specific strategy be assigned to a specific individual and given a projected time frame for completion. Without specific assignments of responsibility and corresponding time frames, the safety committee has no way to effectively monitor the progress of the implementation process.

The following example is provided to illustrate how a completed entry in the improvement plan might actually look:

REINFORCE SAFETY-FIRST ATTITUDES AND PRACTICES

1. Strengthen the safety aspects of our organization's training program for supervisors—especially the part where supervisors learn about how to choose personnel for recognition awards. Assigned to Betty Star, director of human resources. Projected completion date: August 15.

2. Conduct one-on-one conferences with supervisors who have recommended poor safety performers for performance awards. Assigned to John Perkins, safety director. Projected completion date: July 20.

The fact that Betty Star, director of human resources, has been assigned responsibility for strengthening the organization's training program for supervisors does not necessarily mean she will complete all the necessary tasks herself. She might choose to delegate some of the work to subordinates. However, she is the individual responsible and accountable for ensuring that progress is made and that the assignment is completed on time. She is also responsible for maintaining regular contact with the safety committee, giving progress reports, informing key personnel of any roadblocks she encounters, and coordinating any efforts necessary to overcome those roadblocks. The same approach applies to John Perkins, the organization's safety director, and to anyone else given an assignment as part of implementing the improvement plan.

In this way, your organization's improvement plan can be implemented effectively, progress can be monitored regularly, and the necessary human and financial resources can be committed to continually improving the quality of the work environment. This is a process that never ends in organizations that have a safety-first corporate culture, which is one of the reasons they are high-performance organizations—a must in today's hypercompetitive global marketplace.

Nordstrom is one of the most recognized upscale retailers in the United States. With hundreds of stores nationwide, Nordstrom is one of the largest retailers in the world. A retail store is not typically thought of as a high-hazard environment. However, the suppliers worldwide, who manufacture the products sold in Nordstrom's stores, can be. Consequently, Nordstrom maintains a strict set of safety and health guidelines for its supplier partners.

The following statement comes from Nordstrom's Partnership Guidelines: "Nordstrom seeks Partners who provide written standards for safe and healthy work environments and the prevention of accident and injury to the health of their workers, including adequate facilities and protections from exposure to hazardous conditions or materials. These provisions must include safe and healthy conditions for dormitories and residential facilities, and they must comply with local health and safety laws and regulations."

Nordstrom's strategy of extending the safety-related aspects of its corporate culture beyond the doors of its retail outlets illustrates an important point for safety professionals working in today's hypercompetitive, potentially litigious business environment. The point is that responsible organizations know that safety does not stop at the door of their facility. Rather, it must extend beyond to include suppliers, vendors, contractors, and customers.

Source: http://about.nordstrom.com/aboutus/guidelines/default.asp

COLLEAGUE-TO-COLLEAGUE DISCUSSION CASES

CASE 1: Deciding on Cutoff Scores

The safety committee at Jones Corporation has gotten bogged down on a point of contention that has divided the committee into warring camps. One group thinks the cutoff score for determining *acceptable* versus *unacceptable* performance when analyzing the survey results should be the highest possible score. According to the spokesperson for this group, "Anything less than perfect is unacceptable." He and his supporters want every criterion that did not receive a perfect score to be included in the improvement plan. This group calls itself the "Zero Tolerancers." The other group wants the cutoff score to be 3.5 or half a point higher than "somewhat true." This group calls itself the "Realists."

The Zero Tolerancers argue that to set the acceptable performance score at anything less than perfect is to accept hazardous conditions. The Realists counter that setting the cutoff score at 3.5 will give the improvement plan and process a better opportunity to win the approval of higher management. They argue that expecting perfection will simply overwhelm personnel at all levels and cause management, supervisors, and employees to simply lose interest in the safety-first concept. They want to begin at what they think is a reasonable level of expectation and gradually raise the acceptable score over time. The chair of the safety committee does not know what to do. On the one hand, he agrees with the Zero Tolerancers in that until all scores are perfect, the organization's performance is not acceptable, and on the other hand, he thinks the approach recommended by the Realists has more practical merit.

Discussion Questions

1. Do you and your colleagues see any merit in the approach recommended by the Zero Tolerancers? Do you see any problems with this approach? Do you and your colleagues see any merit in the approach recommended by the Realists? Do you see any problems with this approach?

2. If the safety committee chair at Jones Corporation asked your advice, what advice would you give concerning how to proceed?

CASE 2: How Should We Develop the Improvement Plan?

The survey Gail Jones and her colleagues on the safety committee of Scarfron, Inc., conducted recently yielded clear results. Scarfron has a lot of work to do if it is going to have a safety-first corporate culture. Jones and the other members of the safety committee know what needs to be done. What is causing them concern is that the results of the survey might step on the toes of the very people whose support will be needed during implementation of the improvement plan.

As she sees it, several key supervisors and one executive may not like what the survey results indicate about their areas of responsibility. Nonetheless, Jones and the other members of the safety committee know they have to move forward with developing a formal improvement plan and assigning responsibility to specific individuals for specific improvements. What they are trying to decide is how to go about it.

Discussion Questions

1. Have you or any of your colleagues ever had to handle a situation in which there were safety-related problems in the areas of responsibility of uncooperative decision makers, whose support was needed to make the necessary improvements? How did you handle this situation?

2. If Gail Jones approached you for advice concerning how to proceed, what advice would you give her?

Key Terms and Concepts

Before leaving this chapter, make sure you understand the following key terms and concepts and can accurately explain them to people who are not safety professionals.

Assessment	Focus group
Why factor	Strawman
What factor	Survey
Whom factor	Assessment criteria
When factor	Assessment instrument
Interview	Rating scale

Review Questions

Before leaving this chapter, make sure you can accurately and comprehensively, but succinctly, answer the following questions:

1. Discuss the best response to the question, "Why are we conducting this assessment?"
2. Explain the various components of the "whom factor" of an assessment.
3. List four rules of thumb related to the "when factor" of an assessment.
4. Explain the benefits and weaknesses of the following assessment methods: interview, focus group, and survey.
5. Explain how you would develop and distribute a survey instrument such as the one in Figure 10.1.
6. Explain how you would compile the results of an organization-wide survey based on the instrument in Figure 10.1.
7. Explain how you would develop, implement, and monitor an improvement plan based on the results of an organization-wide survey developed from an instrument such as the one in Figure 10.1.

Application Project

Develop a step-by-step plan for conducting an organization-wide assessment of the safety-related aspects of your organization's corporate culture. Make sure the plan answers the following questions:

1. What is your rationale for conducting the assessment?
2. How will the following assessment methods be used: interviews, focus groups, and a survey?
3. How will you distribute the survey instrument?
4. How will you compile and analyze the survey results?
5. How will you arrive at cutoff points for "acceptable" mean scores?
6. How will you prioritize the weaknesses revealed by the survey results?

Endnote

1. Marilyn Wolf Schwartz. "Needs Assessment Pointers," retrieved February 11, 2008, from http://nnlm.gov/psr/lat/v10n5/pointers.html

Index